CONSCIOUSNESS, BIOFEEDBACK AND NEUROTECHNOLOGY.

Raul Valverde PhD.

Contents

1 INTRODUCTION

The trend to explain consciousness by applying quantum theories has gained popularity in recent years and, although clearly disdained by neuroscientists, more and more researchers direct their steps this way up. Brian D. Josephson (1962) of the University of Cambridge, winner of the 1973 Nobel Prize in Physics for his studies on the quantum effects in superconductors (Josephson effect), proposes a unified field theory, quantum nature, that would explain not only the consciousness and its attributes, but also all the phenomenology observed to date in terms of parapsychological and mystical experiences.

Another character that has stood in defence of a theory quantum of consciousness was the physicist Roger Penrose (1994). Penrose, based on the mathematical theorem of Gödel and based on subsequent elaborations, concludes that no system is deterministic; a system that is based on rules and deductions cannot explain the creative powers of the mind and judgment. This nullifies the claim of classic physics to structure deterministic processes into a complex phenomenon of consciousness. Penrose says that only the peculiar characteristics of non-deterministic quantum physics could issue an approximate judgment on consciousness, within a theory that involved quantum phenomena, macro physical and conditions of non locality.

Every human being acts contacting the physical world through his mind, which acts as consciousness. The physical brain is a computer with its data warehouse. The mind is the result of energy which acts to encourage the operation of synchronizing the mental spheres, which are the areas of energy that influence on different levels of consciousness. This energy manifests itself in different ways in order to transform the same energy vibration changes depending on the characteristics of the vibration that is the representation of that energy. The perception of the human energy depends on the direction through which he can capture the vibration it has at the moment.

Biofeedback is the process by which a person learns to influence involuntary body processes to receive physiological data from an electronic device that continuously monitors certain physiological parameters. It is a way of measuring the response to the physical, emotional, mental and spiritual stresses of life. Bodies under high stress are more prone to physical discomfort and even illness. The biofeedback response occurs when the body receives new information about their status (ie, get 'feedback') and make healthy adjustments to reduce stress and tension. The result is a reduction of the nervous activity and increased vitality. Users of the feedback report a greater sense of well-being and joy.

Biofeedback instruments measure muscle activity, skin temperature, electro-dermal activity (sweat gland activity), respiration, heart rate, heart rate variability, blood pressure, brain electrical activity and blood flow. There are many types of biofeedback: GSR, EEG, EMG, CT, MRI, etc. These technologies are able to capture analog electrical signals from the body and translate those signals into meaningful information through complex algorithmic software that a technician can then decipher. Research shows that biofeedback, alone and in combination with other therapies behavior is effective for treating a variety of medical and psychological disorders. Biofeedback is currently used by doctors, nurses, psychologists, counselors, physical therapists, occupational therapists and other professionals.

Quantum biofeedback uses sensors to monitor physiological relaxation indicators, as the temperature of the skin and muscle tension. It expands classical biofeedback by using galvanic skin response (GSR) together with modern computer technology to detect the response of the built-mind-spirit body (sometimes called the super-conscious) to a large array of stress indicators.

Quantum Biofeedback is a sophisticated biofeedback technique based on advanced quantum scientific principles that promote a deeper understanding of health and healing. Seeing the body as a complex electrical circuit and the application of the principles of electrical engineering, who are able to analyze and balance the body electric.

Quantum Biofeedback interacts with all cellular matrix of the body, which allows communication with the conscious and unconscious reactions of the body. This is accomplished by establishing electrical communication using 12 electrodes are attached to the head, wrists, and ankles of the patient. A loop and cyber handshake between Quantum Biofeedback System and the patient using a calibration procedure is then created. By connecting in this way all 200 billion cells in the human body are included in the the exchange of information. On this basis, it is now possible to identify and prioritize the tensions and toxins that are a major concern and bring them into balance.

Classic Biofeedback is based on electrical measurements taken from the front (frontal cortex). When this information is presented to the patient, he tries to consciously change their internal reactions to modify electrical results. Quantum Biofeedback uses our link to all body cell matrix interface, not only the conscious aspects, but also with the unconscious. This provides greater gain in the understanding of health, as our conscious perception of the world is limited by design to 17% of the stimulation we receive.

Monroe (1982), who is considered the creator of neurotechnology, proposed the method HEMI-SYNC: (Synchronization of the cerebral hemispheres by means of sounds) for psychotherapy. This principle states that when a pure tone is emitted, the brain resonates when it receives certain frequencies of waves and synchronized with these, this effect is known as FFR (Frequency Following Response).

Machines created based on FFR began to be popular in the 80s, the typical machine based on the principle of Monroe using stereo headphones that are used separately to send sound signals to each ear, for example 2 signals of 300 and 304 Hz, In one ear will be hear only 300 Hz signal and the other only 304, but since the sounds are combined in the brain, this will hear a third signal of 4 Hz which is the difference between the two sound impulses. This third sign is not an audible sound but an electrical signal that can only be created by the cerebral hemispheres acting in unison and may go unnoticed, producing as a result, that the two hemispheres are focused simultaneously on the same state of consciousness and thus increasing the brain power and inducing it to a different state of consciousness.

The Monroe Institute was created based on the principles of Monroe (1982). The neurotechnology of the Monroe Institute is a system that mixes sequences of sound patterns designed to evoke beneficial brainwave states for different human states of consciousness. Neurotechnology is typically used to tune brainwaves into any range brainwave. With these machines, you may experience theta, alpha, delta waves or including combinations of ranges using frequencies layered that mix several ranges brainwave in a brainwave pattern synergistic as pattern brainwave. Theta brainwaves have been associated with altered state of consciousness by many research studies (Vaitl et al 2005).

An application of quantum physics and neurotechnology in the altered states of consciousness was carried out by Persinger (1983). Persinger (1983) was able to induce visions of God and other religious and mystical experiences in the laboratory by using a computer and with what he called the Koran helmet that has also the nickname of the God helmet. He found that the temporal lobes of the brain are the source of the most spiritual experiences and other alternate state of consciousness.

This God helmet is a device to induce altered states of consciousness using weak magnetic signals to direct brain activity that can induce spiritual experiences that helps with transpersonal exploration and transformation of many people that is able to heal because of these transpersonal experiences (Persinger 2000). It is suggested that altered states of consciousness such as sensed presence and out-of-body experience whether they are produced by magnetic, electric or other stimulations or circumstances can be most effectively explained as the changes of the relative contents and/or intensities of the test subjects' neural quantum entanglement with their surroundings etc. including possibly spiritual environments or information (Hu & Wu 2012).

Persinger and al (2010) reasoned that: "Our primary assumption is that consciousness and its variants of mystical states can be expressed as quantum phenomena". If consciousness and thought are coupled to electron movements, then a macroscopic manifestation should be congruent with the magnetic field strengths associated with neurocognitive activities. Access to the information within the movements of an electron, its fundamental charge, and the photon emissions associated with changes in electron movements, would allow mysticalstates and the information with which they are associated to have alternative interpretations that recruit the fundamental properties of space-time and matter.

With the aid of quantum biofeedback and neurotechnology, people can discover hidden deep events and conflicts in their subconscious that can lead to healing experiences.

2 BRAIN WAVES

Our brain works primarily with bioelectrical energy. Although the power of electricity that handle our neurons is low (measured in mill volts), this power will only have to process, manage, distribute and use vast amounts of information and generate multiple answers (almost infinite in possibilities). So by using micro electricity, we can conclude that the brain is a machine of low frequencies. Our brain, in addition to process information that comes by the senses, is capable of emitting extra-sensory information received via similar "electromagnetic waves", but with lower intensity similar to the frequencies of a radio transmitter-receiver. Our brain can act as a radio station, similar to numerous species of birds that in their migrations are guided by a genuine receiver of terrestrial magnetism located in the pituitary gland. With this receiver, they know where they have to fly and in what direction. It's like having a real compass incorporated into the brain. The mind could be defined as the "sense" of the brain (like sight is the sense of the eye).

A German psychiatrist, professor at the University of Vienna named Hans Berger (1.843-1941), demonstrated that a device "amplifier" that was christened electroencephalograph existed an electrical potential (voltage fluctuations) in the human brain. Before him, the English physician Richard Caton (1842-1926) showed similar potential in dogs. The first types of frequency that were discovered were the "alpha" and "theta".

Later they are complemented by research completed registration range electroencephalograph. Each type of wave results in a different neuropsychological state. That is, our mind, our body and our physical and physiological activity are completely different in each of these states or frequencies. The type of neuro-chemicals and hormones substances discharged into the blood flow varies depending on frequency and as much as the presence and quantity of such substances as the mood we have, interact to produce a physiological-mental-physical-end involves mental state. A level of consciousness is conscious thought that is always thinking even in moments of apparent inactivity, but trivial ideas continually pass by consciousness. Most of the thoughts are accompanied by reactions in behavior, and small involuntary movements. When we are not actively thinking then we are probably dreaming, according Signer and Streiner (1966) a person has throughout the day about 200 daydreams. Therefore consciousness is the sum of all the different perceptions. States are the most common consciousness of wakefulness and sleep;

however, changes in expressing both cerebral and psycho states change according to conscious or subconscious feelings of each person are distinguished. These changes are directly related to the electrical activity of the brain. This activity can be measured by the number of oscillations per second (Hz) and different states of consciousness in the brain: our brain only perceives a limited range of frequencies indispensable to operate with ease in this three-dimensional medium. 20 to 20,000 vibrations per second are perceptible by our ears, the colors perceived by our eyes range from red to violet (although extending beyond, up and down), all possible smells and tastes (which are also vibrations) and the endless textures that we can distinguish with our skin. But the brain is not only receiver but also is sends vibrations. It has been proven thanks to the EEG that the brain emits waves of varying intensity and frequency depending on the mental state of the person being observed. These waves are classified into (Table 1):

Table 1 Types of Brainwaves

TYPES OF BRAIN WAVES	STATES OF CONSCIOUSNESS
BETA WAVES: 14 Hz to 30 Hz	This type of waves are recorded when the person is awake in a state of normal activity. Correspond to states of conscious attention, anxiety, surprise, fear, stress.
GAMMA WAVES : 25 and 100 Hz	They express pathological conditions of maximum tension, excitement and the individual enters a state of STRESS in which the coordination of ideas and normal physical activity are seriously altered.
ALPHA WAVES: 8 Hz to 13 Hz	Relaxation and rest, calm, reflective state. Reduction of bodily sensations. The subconscious begins to emerge: Abstraction, suggestibility. Assimilation of the study. Ease of visualization of mental images.
THETA WAVES: 3.5 Hz to 7 Hz	During sleep or in deep meditation, autogenous training, hypnosis, yoga (whenever the formations of

	the subconscious act). The state stimulates creative inspiration. Considered a state for maximum capacity of learning. Fantasy, imagination. Hypnagogic images.
DELTA WAVES: 1 Hz to 3 Hz	It arises mainly in the states of deep sleep and unconsciousness. Very rarely can be experienced being awake unless with a very hard training (Yoga, Meditation, Zen, Hypnosis, Self hypnosis) or with a synchronizer of hemispheres. It corresponds to deep sleep, hypnotic trance, REM sleep. It corresponds to sleep without dream, trance, deep hypnosis. Delta waves are very important in the healing process and strengthening the immune system.

The alpha state is when, if we would connect to an electro encephalogram, our brains would work between 7.5 to 14 cycles per second. When "operate" in this state, we rank among the border of the conscious and the unconscious. Is the gate between both states of consciousness that makes the "Alpha state" as it allows us, on the one hand, to still have a conscious activity of our mind and body, that is, we realize what we think and happening around us and on the other hand, since the "lift" of consciousness in the "lower floors" of the mind, gives us access to information and own subconscious states, this can use all the wealth of the 90% of "power" mental residing in the subconscious. This state is a "being comfortable" in inner peace and happiness with in deep relaxation. This last state; greatly increases the power of suggestion and auto-suggestion. This is, anything that we suggest alpha, will be more easily accepted with less "filters" by our conscious rationality.

Music is a very effective instrument of suggestion and helps with the production of an alpha state made by many instruments that produce vibrations at that frequency. Hence, with relative ease, many people stay asleep at concerts or simply feel "moved" by the music to another dimension. If we put them to an electroencephalograph, most of those attending a concert in an auditorium would be running on alpha. The contemplation of the sea, the rolling waves and the sound they produce is a good inducer of alpha state. The tone of individuals things like music, prayer, smells, certain types of images (especially of nature, certain types of painting, Tibetan mandalas symbols or own) can be used to put ourselves in this special state of the mind, that makes us touch our innermost and reach the "spiritual zone" of human beings. Therefore, as we see, there are many things that can induce this state.

In 2006, an experiment was designed with the hypothesis that alpha brain wave stimulation would ease the pain of patients undergoing an endoscopy. Forty consecutive patients (25 men and 15 women) were included in the study. Twenty of the patients received photic 9 Hz alpha stimulation for 25 minutes, in addition to the usual premedications. The other twenty patients (the control group) received the same treatment, but without photic stimulation. All of the patients used a five-grade scale to evaluate the discomfort/pain they felt during endoscopy, in comparison with what they had experienced in their previous examination (Nomura et al 2006).

Of the patients who received the alpha stimulation, 18 out of 20 reported feeling less discomfort/pain than they had experienced before, compared to just 3/20 in the control group. Overall comparison of pain scores between both groups was statistically significant (Nomura et al 2006).

The EEG activity of all of the participants was measured during the test, and the group that received stimulation demonstrated significantly higher levels of slow-wave alpha activity. A clear correlation was found in the review of this EEG data: more alpha brain wave activity meant less pain (Nomura et al 2006).

In 2000, a study was conducted among employees at a Dutch addiction care center (Ossebaard 2000) to investigate the possible effects of alpha brain wave stimulation on stress and anxiety.

Subjects in the experiment were given a single stimulation session using 5 minutes of 30 Hz stimulation followed by 35 minutes of 10 Hz stimulation. Before and after the session, all of the subjects completed Spielberger's State-Trait Anxiety Inventory (STAI) test, which is an evaluation tool that is very widely used to determine stress and anxiety levels.

Those who received the stimulation showed a significant, immediate decrease in state anxiety after the sessions, and this effect was consistently demonstrated across 4 tests- the alpha stimulation resulted in lower stress levels every time.

In March of 2006, a paper was published in BMC Neuroscience (Williams et al 2006) which outlined the effects that alpha brain wave stimulation had on a difficult memory task, showing very promising results.

The authors of this paper explain that their study was based on research correlating 10 Hz alpha wave rhythms with memory performance, and additional studies showing that alpha activity declines along with memory due to aging.

During the test, the patients were presented with three letter words, and were given brief bursts of alpha photic stimulation, at various frequencies, as the words were presented. Later, the participants recognition of the words was tested, without any stimulation involved.

The result was that the older participants who received photic stimulation around 10.2 Hz were able to perform just as well on the test as the much younger participants. Those with the same stimulation frequencies increased recognition of the words more than other frequencies and the control, regardless of age.

That work was actually a continuation of a study done by Dr. Williams (2001). Positive results were found during that earlier study as well, where the conclusion was drawn that "10Hz flicker enhances memory in healthy people and may have therapeutic potential in memory disorders."

Alpha brain wave activity has been observed and noted by many studies as an observable indicator of the mental state that many meditation practices seek to achieve. A review paper published in 2006 (Cahn & Polich 2006) cites 19 studies wherein "Alpha power increases are observed when meditators are evaluated during meditating compared with control conditions". The same review goes on to describe an additional 10 studies wherein it was shown that the alpha band is stronger at rest in meditators compared with non meditator controls. The experiments cited include observations of EEG activity from practitioners of many different meditation traditions, including Transcendental Meditation, Zen, Yoga, Tibetan Buddhists practices and Qigong.

Magnus & Van der Holst (1987) speak of frequencies whose frequency ranges from 3.5 to 7.5 cycles per second. They are associated with the states of creative and decisive hallucination. During its broadcast, the person can discover events deeply hidden in the recesses of your subconscious, as well as conflicts and personality in particular. Is the wave of imagination and dynamic contact with other possible "dimensions psi" offering us the opportunity to research our personality and deep in our psychology. Through it, the person may discover their deepest secrets. This frequency is "issued" by the disciples of the Zen philosophy, Buddhist monks and Christian religious in their meditations, songs or prayers. It is primarily generated unconsciously during the period preceding the nocturnal sleep and wake just before and certain hypnotic states.

Gamma brain waves are the highest frequency brain wave type. A variety of studies have associated gamma with the formation of ideas, linguistic processing and various types of learning. Gamma waves have also been linked to the cognitive act of processing memories- the rate of the waves seems to correlate with the speed at which a subject can recall memories; the faster the waves, the faster the recollection.

Gamma waves have been shown to disappear during deep sleep induced by anesthesia, but return with the transition back to a wakeful state. A recent Scientific American article discussed gamma waves in conjunction with long-term Buddhist meditation practitioners. In this article, it was found that the experienced meditators demonstrated self-induced, high-amplitude gamma oscillations during meditation. Researchers also noted that their gamma activity differed significantly from those in a control group, both during the meditation and before they even began. Interestingly enough, a similarly strong presence of gamma waves throughout the cortex has been observed in musicians listening to music, compared against a control group of non-musicians.

One current theory even suggests that gamma brain waves may play a role in creating the unity of conscious perception. Research into this theory is still ongoing, and the question is a difficult one to answer with certainty at this time. Though a further investigation is yet to be completed, the theory points to a very interesting possibility that gamma waves are involved in self-awareness (Huang & Charyton 2008).

Gamma waves have been used for Cognitive Enhancement. In a study lead, 30 students, ranging in age from 6 to 16, the students were given a 35 minutes of gamma wave stimulation twice a week for 6 weeks. Participants put on headsets and light goggles and completed each session while reclining in a comfortable chair.

Before and after the 6 weeks of treatment, each participant was tested using the Wechsler Intelligence Scale for Children, Third Edition (WISC-III). Analysis of the test scores from before and after the gamma sessions showed that the 30 participants demonstrated significant progress in a wide variety of tests measuring cognitive abilities.

Researchers found a statistically significant gain in the participant's speed of information processing and visual motor coordination on the Symbol Search subtest (pre-test mean: 6.9, post-test mean: 10.6). There was also a statistically significant gain in the participant's visual short-term memory and sequencing ability, as measured by the Coding subtest (pre-test mean: 6.0, post-test mean: 8.2). The Arithmetic subtest also revealed significant improvements (pre-test mean: 6.2, post-test mean: 8.3), demonstrating a significant gain in the students number ability and short-term memory. Freedom from Distractibility, which is a measurement of ability to focus and pay attention, increased significantly (pre-test mean: 13.2, post-test mean: 17.5) as did Processing Speed (pre-test mean: 12.9, post-test mean: 18.8). The authors noted that the relatively low number of AVS sessions needed to improve cognitive abilities served as a further demonstration its efficacy.

Gamma waves have also used for treatment of migraines, a study was conducted by David Noton (2000), with the goal of further validating the use of brain wave stimulation using light as an aid for migraine sufferers. Dr. Noton mentioned in his study that the treatment of migraine with light-based stimulation originated in the late 1980s with the work of Dr. Duncan Anderson, a neurologist at the Royal Postgraduate Medical School at Hammersmith Hospital in London. He found that when a patient was presented with red light flickering at 30 cycles per second for 15 minutes, the patient experienced a significant reduction in the frequency of migraine attacks, and occasionally would experience nearly complete remission of their migraines. The same treatment, applied for up to 30 minutes after the start of a headache, sometimes terminated the attack.

Accordingly, subjects in Dr. Noton's study were asked to use photic gamma stimulation for 15 minutes per day, and for at least 30 days. After 30 days of daily use, all of the participants were given a survey. Out of the total of 55 regular migraine sufferers, 44% reported that the frequency of their migraine attacks after the treatment was either 'Somewhat Less' or 'Much Less' (under a conservative interpretation of these categories). Within the group of 28 migraine sufferers who stated that their migraine attacks were normally preceded by warning signs, 53% reported that the frequency of their migraine attacks was 'Somewhat Less' or 'Much Less.'

The study concluded that "in view of the limited efficacy and undesirable side-effects of the available migraine preventive drugs, photic stimulation (flickering light therapy) must be considered a possible preventive treatment for migraine."

Beta is generally the mental state most people are in during the day, and usually this state in itself is uneventful. However, beta brain wave activity is significant to proper mental functioning, and insufficient beta activity can cause mental or emotional disorders such as depression, ADD and insomnia. Broadly speaking, beta brain waves are associated with alert attentiveness and concentration- intense focus and problem solving are linked to beta activity. Beta waves can also be related to strong, excited emotions. Medications that are designed to induce concentration and alertness, such as Ritalin or Adderall, actually produce a beta brain wave state in most subjects (Huang & Charyton 2008).

One of the many examples of research connecting brain wave stimulation to increased cognitive abilities was conducted by Joyce & Siever (2000), and focused on the reading and verbal skills in elementary school students in Minnesota.

Before and after the study, the students were tested on the STAR (Standardized Test for the Assessment of Reading). The experiment consisted of 31 sessions across 7 weeks. For the first 8 sessions, 7-9 hz alpha brain wave stimulation was used. The remaining sessions consisted of isochronic tone stimulation, targeting high alpha in one ear, and beta ranges in the other. After the 7 weeks of stimulation, the students improved their reading level by over half a grade- a .6 GE improvement. By comparison, the control group only improved by .2 GE. The grade equivalent (GE) score ranges from grade 0 to 13 and represents a student's actual grade reading level. For example, a student with a GE of 4.7 is reading at the level of a typical student in the 7th month of the 4th grade. Another measurement used in this test, the percentile rank (PR), ranges from 1 to 99 and indicates a student's reading/verbal ability compared to his/her peers nationally. In this measure, the control group's performance actually decreased slightly, by 1.2. However, the experimental group improved considerably, raising their average PR by 7.2.

In addition to their academic improvements, teachers and parents noted general social improvements, as documented by the research team. One teacher commented, "during the time students participated in the AVE (audio-visual entrainment) program, behavior began to change. Brains were engaged, and students were more alert. And they began actively participating in the teacher-student dialogue portion of the class". Another significant milestone in the history of brainwave stimulation for cognitive enhancement can be found in by Budzynski et al. (1999).

This study was designed to test the hypothesis that extended beta brain wave stimulation would result in positive changes in academic performance. The participants in this study were university students who had signed up for academic counseling, and they were divided into two groups of 8 each. The experimental group received 30 sessions of 14-22hz beta stimulation. The control group received no training.

The training consisted of 15 minute sessions, 5 days of the week for 6 weeks. Subjects were allowed to adjust light intensity levels, based on earlier work finding great individual differences in preference.

To prove that academic performance was enhanced, the group who received stimulation would have to have shown a significantly greater increase in GPA than the control group. That is exactly what happened: students who received the beta stimulation increased their GPAs by a statistically significant degree in the semester after the training- by .62 points on average. On the other hand, the average GPA of students in the control group decreased by .22 points in the following semester.

Another interesting finding in this study is that direct connections were found between EEG activity and student performance. For example, significant correlations were observed between 12-15hz beta activity, and performance on the Digit Span and Symbol tests, as well as 12-15hz beta band activity during arithmetic performance. The subjects who performed best on these tests were demonstrating higher levels of beta activity.

Howard (1986) conducted a study with the aim of finding methods to reduce exhaustion and stress among dental students. He noted that the negative impact of the heavy workload for these students was well documented and of concern. He conducted a study using photic and audio beta brain wave stimulation, with the supposition that the therapy may reduce fatigue and stress.

The group receiving the treatment included 12 dental students, who completed a course of seven 30-minute sessions over a seven-week period.

The students who received beta wave stimulation experienced significantly lowered levels of fatigue by the end of the seven weeks, as measured by the POMs (Profile of Mood States) test.

Beyond that, students receiving the stimulation showed notable improvements in the quality of sleep they were able to attain. The number of days during which they experienced a sleep disturbance was reduced by more than half, from an average 4.6 down to 2.2. The number of disturbances on a given night and the severity of the sleep interruptions were also greatly reduced.

Theta brain waves (3-8hz) have been connected via extensive study to many different phenomena in the brain. Research on subjects as diverse as memory, emotion, neural plasticity, sleep, meditation and hypnosis have all drawn links to theta activity. A theta state is associated with stage 1 sleep- very light sleep from which subjects can easily be awoken.

Connections between meditation and theta activity have been researched and documented thoroughly, particularly in the case of both Zen and Transcendental Meditation. Meditative theta states are often associated with vivid mental imagery, peacefulness and generally pleasant experiences.

More recent research highlights the interesting role that theta may play in memory function. One theory proposed by Lisman and Idiart suggests that short term memories are constantly refreshed in order to retain their presence in the brain while they are being accessed. They suggest that individual memories are refreshed at the gamma rate, while the whole refresh cycle pulses at a theta rate. They believe that this may be why an average of 7 items can be held in short term memory by most people - per each 6Hz theta cycle, the 40hz gamma can cycle an average of 7 times (Lisman & Idiart 1995).

On the significance of theta brain waves, Gabe Turow (2005) writes: "The links between the theta frequency and memory, emotion, and neural plasticity on a localized level provide relevant clues to questions on why visualizations of meditators in theta are so vivid, why meditators have such good memories, and why hypnosis can create lasting changes in the brain.".

In 2001, a study was conducted with the objective of determining "whether mildly anxious people would report decreased anxiety after listening daily for 1 month to tapes embedded with tones that create binaural beats." (Le Scouranec et al 2001).

The experiment was performed with a volunteer sample of 15 mildly anxious patients. Participants were asked to listen to the tapes at least 5 times a week for 4 weeks, and to record tape usage, anxiety ratings, and other comments in a journal during the study.

The most frequent comments included in patient journals were about enhanced relaxation and falling asleep. Many indicated that as the study progressed, the relaxing effects of the simulation began earlier during a session. They reported that being relaxed helped them to experience sleep onset faster, to sleep better, and to awake more rested. Many comments indicated that participants had a hard time staying awake when listening to the tapes, and they sometimes reported that they fell asleep with the tape still running.

Participants were asked to rate their anxiety on a scale of 1 to 100, with 1 indicating no anxiety, and 100 marking very high anxiety. Before listening to the tapes, the average score reported was 43.1. After listening, the average dropped significantly, to 24.3.

The study's authors concluded: "Listening to binaural beat tapes in the delta/theta electroencephalogram range may be beneficial in reducing mild anxiety."

William S. Kroger (1959), in his influential 1959 study regarding hypnotic induction, began the article with the following observation:

"In the year 1784, Benjamin Franklin, a pioneer in electricity, denounced as a fraud, Mesmer, a pioneer in hypnosis. Today, 175 years later, the work of one is aiding the work of the other as electronic devices induce the hypnotic state."

Kroger and Schneider tested the Brainwave Synchronizer on 2,500 patients before writing their report. When their data from all the tests was combined, they found that the photic stimulation provided by the device was able to induce a hypnotic state in just over 5 minutes, for nearly 80% of subjects- and of those subjects, the percentage of deep hypnosis inductions was twice that of those who remained in light hypnosis.

"Expectation level" was found to play a role in the success of hypnotic induction produced by the photic stimulation. Subjects who entered the experiment "cold," with no previous explanation regarding the device and hypnosis, were less likely to be inducted. But "when the instrument was used as the final part of a program containing several demonstrations of hypnosis in the conventional manner and a detailed explanation of what the instrument will do" the hypnosis induction success rate reached reached nearly 90%.

However, prior experience with being hypnotized was not found to be a significant factor in this study. The authors kept separate figures for groups with and without previous hypnotic experience, and found only a negligible difference in the results between the two.

Since that 1959 study, ongoing research has further solidified the connection between brainwave stimulation, theta brain wave activity, and ease of attaining a state of hypnosis.

With two subjects who were minimally responsive to hypnosis, Wickramasekera (1977) found that after 10 sessions of deep theta stimulation, including suggestions encouraging hypnotic responsiveness, the subjects increased an average of 6.5 points on a 12-point hypnotizability scale.

Sabourin et al. (1990) found that, "in eyes open and closed conditions in waking and hypnosis, highly hypnotizable subjects generated substantially more mean theta power than did low hypnotizable subjects at all occipital, central and frontal locations in almost all conditions of waking and hypnosis.".

Observations of the connection between increased theta activity and meditation practice go all the way back to 1966 (Kasamatsu 1966). Cahn (2006) lists 29 individual studies showing a correlation between theta activity and active practice of various meditative traditions- primarily Zen and Transcendental Meditation.

One of the more recent studies (Takahashi et. al, 2005) monitored EEG activity in 20 adult Zen practitioners. This study specifically utilized "So-soku," a Zen meditation practice for concentrating the mind by slowly counting one's breaths. In Su-soku, the subject starts counting 'one' when they exhale, and inhale naturally without counting. The counting continues until the subject reaches 'one hundred' and the subject then starts again from 'one.' If other thoughts occur, the subject lets them pass and restricts their attention to counting.

As a control condition, participants were asked to simply try to keep their breathing regular for 15 minutes, and EEG data was monitored for that time. Afterwards, participants left the room for a 30 minute rest period, before returning to perform the So-soku meditation for 15 minutes.

When the EEG data collected during the respective 15 minute periods was analyzed, significantly higher theta power was detected during the meditation practice vs. the control condition.

Personality trait factors were also analyzed during this experiment. And it was found that the percent change in fast theta power in the frontal area, reflecting enhanced mindfulness, was positively correlated with the harm avoidance score. As the authors noted, this has been suggested to be associated with increased serotonin activity.

Delta waves are the slowest of all brain waves, and are predominantly associated with Stage 3 and Stage 4 sleep. However, researchers have found functions for delta stimulation beyond those related to sleep, as you have seen in the studies above. Stimulating delta in people who are wide awake has been shown to have additional benefits such as increasing relaxation and relieving certain types of pain (Huang & Charyton 2008).

A 1985 (Solomon, 1985) study provided some of the earliest clinical evidence for the ability of delta wave stimulation to aid in pain relief, and specifically the relief of headaches.

In this study, 15 patients with acute muscle contraction headaches were given just 5 minutes of delta stimulation at 1 to 3hz. All but one of the patients reported complete relief of their headache in that short time.

4 of the patients were also treated in a placebo-controlled trial- none of the 4 responded to the placebo, but all 4 experienced headache relief when given actual delta photic stimulation.

This experiment included an additional 6 patients who were suffering from chronic muscle contraction headaches- and again, all but one patient experienced complete headache relief after 5 minutes of delta stimulation.

The study reached the conclusion that "slow wave photic stimulation appears to be effective in the treatment of acute and chronic muscle-contraction type headaches.".

A study was conducted to explore whether brain stimulation could be useful in reducing preoperative anxiety in patients, while still allowing them to be "street ready" after the operation-meaning their mobility and functionality would not be hindered (Padmanabhan et al 2005).

108 patients who were scheduled to undergo general anaesthesia for elective surgery were recruited for this controlled study. About an hour before their operation, the subjects were asked to complete a State-Trait Anxiety Inventory (STAI) questionnaire. The STAI is a well validated and widely used test for measuring subjective feelings of anxiety. Scores for the STAI range from 20 to 80, with a higher score corresponding to higher anxiety levels.

Subjects were separated into one of three groups. One group listened to a 30 minute audio track including music and binaural beats stimulating delta activity. Another group listened to exactly the same music, but without the binaual beats. The third group received no specific intervention, and participants were allowed to read or watch television.

At the end of the 30 minutes, participants completed the STAI questionnaire for a second time. The group who listened to delta binaural beats experienced the greatest reduction in anxiety scores, a 26.3% drop. By comparison, the group who only listened to music lowered their scores by 11.1% on average, and the "no intervention" group lowered their scores by just 3.8%.

This is one of a number of studies that has successfully shown that the effect of audio brainwave stimulation is not just a placebo, nor is it the same relaxing effect that one may get from simply listening to relaxing music. These carefully executed and peer reviewed studies give evidence for a very distinct benefit specific to this form of therapy.

Three different treatments were tested in the study of Siever (2003), with the aim of finding ways to increase the quality of life for those diagnosed with Fibromyalgia Syndrome.

The 49 participants were randomly distributed into three groups. One group would receive brainwave stimulation. The second would receive standard and alternative medical therapies such as prolotherapy, neural facial therapy, and acupuncture. And the third would receive nutritional supplements, consisting of amino acids, vitamins, minerals, and herbs.

The brainwave stimulation group listened to a 30 minute delta session every night while in bed. The delta stimulation was delivered through a pair of headphones (using isochronic tones), and the participants either removed the headphones when the session ended, or simply fell asleep with the session still playing.

All of the participants completed the SCL-90 survey before treatment, and after one month. When survey scores from all three groups were compared, it was found that more patients from the brainwave stimulation group (58%) experienced improvements in sleep and reductions in fatigue than did patients in any other group.

According to Tornell (2001) When the actual consciousness is reached, the union arise between attention, observation, and deep memory of the upper levels of the inner being is when you wake up the so-called third eye.

Attention is based on vacuum created by thought, and this is very common in sleep, but in conscious state is when one sets the delta waves intention, attention to capture the flow of energy called chi, or prana is a flow of energy of great power differential that creates the quantum vacuum for the big quantum soup in their upper chakras polarize.

That yes and no, they are alternative of awareness, and this occurs in the frontal lobes, between observation (care), and memory. The right hemisphere is biased to capture internal events as dreams based on chaos, theta waves, hence the great attractor (light sleep), but also reaches certain levels of alpha waves.

According to Tornell (2001) in night dreams, you go through four stages of sleep or the four stages of waves and depth of field, sleep is very necessary to restore sanity, precondition to download the emotional tensions affect brain function. This is known as REM sleep.

The occipital brain is related to reflection, and 8-16 hertz cycles, this state occurs in deep reflection or transcendental meditation, at this stage the brain images can replace words equations numbers and perform simulations through the screen thalamic.

At this stage the alpha than 12 cycles, waves occur in people who have gone through initiations or very near death experiences and this corresponds to levels of cosmic consciousness.

The left hemisphere is associated with the senses, and its waves ranging from 16 to 32 cycles. 24 in the 16 cycles as carrier wave, and in the range transversely with these cycles, music stands, and audio, this goes frequencies of 60 cycles to 14,000 cycles, hence the electromagnetic spectrum, from 3,800 to 7,600 light goes from red to violet. Field of the right hemisphere, creates science and experience, while the left creates the vision, revelation and creativity, but the goal is to use all sides of the brain simultaneously and the whole body as an integral whole.

For the vast majority of consciences is not possible greater interconnection of the corpus callosum, reflection and visionary dream alert they are out of reach, in an associative confusion.

The conscious ego is disconnected from the transpersonal inner being, there are many blockages between the hemispheres where one dominates the other and an illusory appearance of artificial symmetry is created, there is no internal consistency that covers everything, including the chaos and unpredictability.

We find ourselves caught in a consensual reality caused by a vision limited to the left hemisphere. The differences between the dreamer and the director must end and allow the two hemispheres to communicate through that brain interface that is the corpus callosum.

The corpus callosum is more than 700 million fibers interconnection and each fiber is capable of transmitting more than 300,000 simultaneous data in both directions, only 07% of these fibers is used, which gives us the ability to increase capacity without to genetically modify the brain.

We fully complete our spinal neuro brain system is capable of simultaneous access to all dimensions, we only need to grow in awareness and the ability to crystallize energy. The communication of the cerebral hemispheres by the corpus callosum and cerebellum is the most important part of the overall integrated communication, but there is another alternate route is the communication between hemispheres through the physical body.

This type of communication is slower because it involves emotional and sentimental brains, as suffixes preprograms established for knee-jerk reactions: automatic and not conscious ego. Many scholars believe that at this point, is where human beings act as an integral whole conscious, but this is not entirely true, at all.

One thing is that the body is involved in the mundane daily activity, and another is the use of awareness, reflection and quantitative, qualitative analysis Attention intention and higher to form a pattern of quantum reality in its entirety consistent element. Another is that the multidimensional energy vortexes located along the column are used as suppliers of energy and information at various levels of quantum line 25 so that the entire brain to make its integral work of intelligent and reflective consciousness.

The consciousness is in the brain, nervous system, limbic, and trunk, also act pre programmed level of instantaneous response (reflex arc), any activity that involves conscious thoughts and mind are given in the brain and its attached peripherals.

In the ancient philosophies of sexual energy it is involved as a major mini brain that provides energy to neurocortex so that it can deploy to reverse the physicality inherent powers.

In the initiation schools of ancient Egypt they knew this, in India and Latin America also, why initiations involve exposing the candidate to situations impending doom.

To cause the system to act as a complete unit by removing the left hemisphere control body coherent reflection of the total, or also known as pure consciousness body.

At that moment the root chakra (prostate), the sexual chakra (gonads), and abdominal (adrenal), are activated and generated an unusual amount of adrenaline and vital chi not going up the spine (sushumna), nerves, and the endocrine system (ida, pingala). To give the brain a lot of energy, neuropeptides, endorphins that activate the sleepy brain.

It is where neural connections are activated to the extent of creating new neural pathways or dendritic synapses, between them, breaking the barrier of lack of corpus callosum energy.

Causing the integration of the two cerebral hemispheres with the cerebellum, so this acted to greater capacity and can deploy certain characteristics (powers), which allow the initiated to resolve the situations of imminent death that was exposed.

To overcome this state of existential crisis existential and actual initiation occurs that called spiritual rebirth, being that he was taken to the state and will never be the same again.

This is because the human body acts as an integral whole, where the cerebrospinal new neuro perform reconnections, and raise others that were not in use, (creates new gap), and then we just have to remember how and cross it again, the Sufi technique is quite dangerous. For a revival experience is liberating and not traumatic negative, it is necessary that the motive cause and effect so I will step otherwise understand this can create more permanent blockages and release instead of jails and blocks. The average human cell has the same number of molecules that cells having the human brain, the fertilized human egg cell is even more number of molecules, because it has to form its own micro universe or human body. In this mitosis pregnancy it occurs 52 times, before forming the basic stem cells.

Liboff (1985) explains that the earth resonant frequencies are responsible for biological, such as menstrual cycles and circadian rhythms and patterns of behavior and emotional. The frequencies are then picked up by the flora and fauna which are biological instruments that respond to the wave patterns wave patterns resonate in the cranial structure of our head and converge in the center of our brain which is where we find the pineal gland. The pineal gland is believed in many cultures to be the spiritual third eye responsible for intuition. Descartes called the seat of the soul call, where mind and body meet each individual cell in our body receives an electromagnetic impulse from our central nervous system. They receive the same impulse that spreads to all biological instruments on earth.

3 BRAIN ACTIVITY

Berger (1929) discovered the encephalogram (EEG). Since then there have been studies and research aimed at the knowledge of the states of consciousness through measuring encephalographic patterns. Unfortunately, the technique has advanced at the pace that everyone would like. Today research has discovered few methods that can be assisted by the computer such as the magnetometer, the TAC (Gliem et al 1976), that is would not measure brain activity but it would show morphologically what is the situation of the brain, which can help us to try to learn a little more about brain structure, as otherwise unknown to humans.

The electroencephalogram measurement is based on some electrical patterns which emits the brain that corresponds to the state which the brain is. The measurement is made by means of electrodes which they are located on the scalp, approximately above the cerebral lobes be they frontal, occipital, temporal or parietal; They are located in both hemispheres said electrodes and a measurement is made, either at rest or states causing hyperventilation or apnea in order to observe EEG patterns that are registered. Like every technique, it has its drawbacks, if the electrodes are placed very close to each other, it can give records that can be very different. Furthermore, the electromagnetic noise and involuntary movements scalp, which is necessary for calibrating the electroencephalogram (EEG) might given results that are not definitive. However, today it is easier to analyze these results since they can be collected via computer and compared with the results of multiple EEGs, then noise is filtered then and the pattern we get is averaged in order to remove possible bias.

Electroencephalograms have defined very well the brain states through its brainwaves (Wise 1995). The brainwave "beta", which is the rate corresponding to the alertness, which ranges between 14 and 22 cycles. The brainwave "alpha", which would be the pattern when we are relaxed, to the stages before sleep, of visual imagery, and that goes from 8-14 cycles. The brainwave 'theta' ranging from 4 to 8 cycles, and the "delta" brainwave that would be 2-3 cycles per second and would correspond with deep sleep.

Measuring electrical activity would be job of the EEG, but also the brain human has a constant magnetic activity and this activity in turn generates a magnetic field around the head. This magnetic field originates on the activity of brain neurons and is of weak intensity, and its production is due to ionic shock that occur between neurons. Neurons receive nerve impulses from other neurons through its branches, dendrites, according to the arrangement of which dominates the magnetic field direction in the cortex, which in this case is perpendicular to the surface. When a nerve impulse reaches the synapse, it is the junction between two neurons, then they release neurotransmitters into the synaptic cleft and this leads to the stream of a sodium and potassium ions, which facilitates the electrical impulse across the synapse and thus allows the pulse to be transmitted. It is believed that brain magnetic fields are mainly due to these postsynaptic currents. The problem is that these magnetic fields are of very low intensity, equivalent to one billionth of Earth's intensity magnetic field (Nunez and Srinivasan 2006).

The technique that measures the intensity of the magnetic field is cerebral magneto encephalography, it is as MEG. The MEG is a completely non invasive technique. Detection of brain magnetic fields is very difficult, because although not distorted the shape of the skull, scalp, etc., the MEG still has clear limitations, especially in regards to its sensitivity and size, because currently they are bulky and very expensive equipment. However, magneto encephalography is ahead of brain research, especially for the study of disorders like Alzheimer's disease, epilepsy, etc. (Hämäläinen et. al 1993).

A significant role in the evolution of MEG was because the Physics Nobel prize Brian Josephson (1962), because of its important work with superconductors, which states that superconductors operating at temperatures extraordinarily low electrical noise is almost zero which increases meter sensitivity.

Another modern device that has been of great importance is the CAT-SCAN, this technique that can accurately describe the topography brain dynamics. We also have PTSD or "Positron Emission Positrons' and NMR, or system of "Nuclear Magnetic Resonance". The CAT-SCAN provides bioelectric information about the brain level as if it were an electroencephalogram, with the variant that the paper is replaced by a screen computer and it is widely used for biofeedback, and its principles are very simple: assign a color to each type of wave. For example, the alpha wave may be red, green beta, etc (Hutchison 1986).CAT-SCAN allows the screen appearance of a colored map of the state of the brain, coloring being indicative of each wave type area. For example: if started a relaxation, we can see how slowly, as the brain immersed in a relaxed state, higher brain areas are colored blue. By contrast, if at a certain point the experienced open your eyes, it is possible to see how slowly the color of beta waves of wakefulness go coloring again brain areas.

The CAT-SCAN is now commonly used to teach driving brain frequencies of its users. Thus, a person who wants to learn to relax can see how your condition affects the brain relaxation techniques used for this: for example, slow breathing, muscle relaxation, progressive, etc., and will be seeing more and more areas of the brain take rather bluish red beta waves, as it enters the state relaxation, and associate it to the blue hue that slowly fills the scheme of your brain on the screen.

The CAT-SCAN also allows us to know the degree of hemispheric synchronicity (More et al 2000). The brain has two hemispheres, right and left, a more analytical and the other more creative. In order to get to ecstasy, to climax of meditation, it must prevail over any hemisphere but both should be synchronized, that is, broadcasting the same pattern waves.

Another device is called "mind mirror" (MM). This is a device capable of detecting both different brain waves through a number of points of light called "leds" which create a series of graphics in the form columns which can then be compared and columns according to these maps resulting brain activity we see the individual connected to the device (Charman 2000). The doctor Maxwell Cade (1989) has been the principal investigator for the MM, to the point which means it can get to capture the different hierarchies of states of consciousness, from waking to relaxation. As the CAT-SCAN uses four primordial frequencies, the MM detects up to sixteen different simultaneous frequencies. Dr. Cade (1989) established a hierarchy of states of consciousness, we can cite as an example some of them:

- High beta in left hemisphere: This means a heightened state of focus and concentration led to the outside, that is, the person is logical.

-Alpha wave symmetrical. This is a normal alpha state, that is, relaxation, although alertness. There is a void of ideas and images.

- Alpha wave: The subject is in a state of passivity. It is detected also a high beta, so the alert does not disappear. They can also appear, in unison, theta and delta activity, so the subject is deeply relaxed. Given its characteristics, some researchers say it gives a strong power of suggestion in this state and therefore also great reliability to capture information.

- Alpha theta symmetrical and down. It is the typical state of the meditation.

- Apha very high range with cerebral timing: It is the state of mind awake, lucid consciousness, or fifth state, as it is also called. It is a state that is accompanied by feelings of euphoria and increased mental faculties. The MM is used for detecting state receptive to intuition, during the process of rapid breathing.

4 CONSCIOUSNESS

Being aware is to realize everything that happens around you. It is as if in the previous moment you take an action, before acting you ask yourself: Is it right what I do? And someone who "is not you" I answer only "yes" or "no". This answer is a rating of the act are going to run on your "own will" (Hawnser 1997).

Now our mind before ordering action to the physical body, is a study of the elements it has to do: knowledge, physical strength, skill, etc. Consider the risks, the consequences, the material benefits that give results us act in a way or another, also loss and damage and / or suffering may bring action ourselves or third parties, and that interest would we have in that harm or benefit.

The spirit of every human being acts contacting physical being through his mind, which acts consciousness. The physical brain is a computer with its data warehouse. The mind is the result of energy which acts to encourage the operation synchronizing the mental spheres, which are the areas of energy influence on different levels of consciousness. The energy manifests itself in different ways. To transform the same energy vibration changes depending on the characteristics of the vibration is the representation of that energy. The perception of the human energy depends on the direction through which can capture the vibration it has at the moment (Hawnser 1997).

Edelman and Tonomi (2000) think that the whole universe is actually one living organism with full conscious awareness of self. The reason why it may seem difficult to comprehend this is because our understanding is typically limited by our way of seeing the world. When the body tend to listen consciously term anthropomorphize its definition, giving it human qualities. Mistakenly we look past what an organism truly is in the first place. The definition of an organism is any living being able to respond to growth stimuli reproduction and development and maintenance of homeostasis in his stable set our universe does. All these things. The consciousness of our universe is responsible for the form and purpose that all matter assumes.

Carl Jung (1981) found that there is a collective unconscious connected to all humans, meaning that all humanity shares a single mind with one another. This is evident in the world through accounts of shared mythology and symbols. This collectivity is a global example of the unconscious mind of the human body in which billions of cells share a similar signal. Human consciousness is an electromagnetic energy field, this could explain many paranormal phenomena as telepathy and clairvoyance.

The universe, nature and creation becomes conscious of itself (self-conscious) through self-reflexive brain and generating state. The union of different brains in one system creates a supra-consciousness or superior knowledge in network system. Individuals who are not only "individual" but also "social" consciousness would be those precursors are called to produce real social change.

Consciousness is defined as sensory perception of sights, sounds and smells that are close to us. But our consciousness includes many more perceptions. Any person through introspection be able to perceive first lot of bodily sensations such as temperature, touch, different pressure points on your body, release tension, the rhythm of your breathing or the beating of his own heart , saliva, the texture of his clothes, itching, pain, etc. Besides these different physical perceptions, through an even deeper concentration will become aware of your mental and emotional state, you perceive the constant variety of thoughts and different emotional states they generate from the joy caused by a memory spontaneous happy moments childhood to sudden sadness at the thought of a tragic event. It will become aware of vague emotions like irritation, excitement or boredom. The person can achieve an even more abstract perception being aware of time, mortality itself or the continuity of its consciousness and individuality of your conscious self. This personal perception of inner experience itself is particularly evident in the moments of greatest emotional intensity however is always present to some extent in current and even bored with our lives now.

Most Western scientists assume that consciousness is produced in some form by the brain. There are of course evidence for that position. There is evidence of common sense in our daily lives. When a person drinks too much alcohol or a hard hit to the head, you do not think clearly. We also have more sophisticated tests of the relationship between the brain and consciousness.

In fact all the theories of consciousness during the last century has been supported by psychologists who have been moving toward materialism that characterized the nineteenth-century physics based on Newton's classical mechanics. These have been trying to show that consciousness is only the functioning of the physical brain. This materialistic psychology was supported by John Watson (1916), who wrote that psychology is a purely objective experimental branch of science that needs no consciousness in the same way that science does not need chemistry and physics. It is ironic that while Watson linking psychology to classical physical knowledge of Newtonian physics this to non-materialistic faced overwhelming experimental evidence that the universe is related to quantum physics that could not be made without reference to consciousness.

John Lorber (1978) specialized in children with hydrocephalus, or water on the brain British neurologist. Children with this condition have an abnormal amount of cerebral spinal fluid accumulation in the cavities inside your brain compressing brain tissue and usually leads to mental retardation seizures, paralysis and blindness if not treated to death. However Lorber describes dozens of children and, finally, some adults with severe hydrocephalus. But it seems to lead a normal life. Indeed, in a sample of children in the cerebral spinal fluid it filled ninety-five percent of its skull leaving virtually no room for any brain tissue. Half of them had a higher IQ than one hundred thirty years Lorber published an article in the prestigious journal Science titled Is it really necessary brain?

The brain scan on the left side of this slide is a normal brain. The gray area is the cerebral cortex of the brain that thinks and black area in the center is the cerebral spinal fluid in the central cavity to the right is the brain of an adult with a severe hydrocephalus. The vast majority of the head is filled with cerebral spinal fluid with only a thin crest pressed against brain tissue. As the skull barely has enough to let this person come to live, let alone function normally in accordance with modern medical neuroscience. However this particular analysis of the brain of the person with hydrocephalus was actually a graduate student in mathematics at the University of Cambridge with an IQ of one hundred twenty six.

Some of the best evidence that consciousness can function independently of the brain come from near death experiences, profound experiences that some people report when they have been on the threshold of death. The near-death experiences are very short stories of people who have been clinically dead and then are resurrected or revived spontaneously after a brief interval with the memory of what they experienced during that period. According to Greyson (2010), people usually reported vivid mental clarity exceptional sensory imagery a clear memory of the experience and an experience that is more real, then in their daily lives. All this occurs under conditions. Brain function has been drastically altered in the materialistic model would say that consciousness is impossible, these near-death experiences are reported by between ten and twenty percent of the people who revived from clinical death. Greyson (2010) investigated nearly a thousand such cases. The average age at the time of the near-death experience was thirty-one years. But there was a very wide range. A young girl reported an experience I had when I was eight months old and undergo kidney surgery. The near-death experience or older have studied was eighty-one at the time of his heart attack. About a third of these near-death experiences occur during surgery. A quarter in the fourth serious illness and accidents happen happens during life-threatening. The common characteristics of near-death experiences can be categorized as changes in thinking changes in emotional state. Paranormal features and characteristics otherworldly changes in thinking during the near-death experience include a sense of time is altered. People often report that time has stood still and ceased to exist during the experience. It also includes a sense of revelation or sudden understanding that everything in the universe suddenly becomes crystal clear.

There was a sense of thoughts of the person who is going much faster than usual and be much lighter than usual. And finally there was a life review or panoramic memory in which the whole life of the person appears to flicker in front of them. Typical emotions reported during near-death experience include an overwhelming sense of peace and a sense of well being, cosmic unity or oneness with all and a feeling of complete joy and a sense of being loved unconditionally. Paranormal features often occur in the near-death experiences are a feeling of leaving the physical body sometimes called an out of body experience and the experience of the person physical senses such as vision and hearing more and more alive than never. Sometimes people report seeing colors and hearing sounds that do not exist in this life and a sense of extra sensory perception know things beyond the range of the physical senses, as things that are happening in a remote location. And finally visions of the future. Finally, many people report that their near-death experiences came another underworld or realm of existence.

Many reports came to a border that could not cross or a point of no return than if they had not be crossed and many reported that were allowed to return to life. Many reports were found to be some kind of mystical or divine and some reports often see spirits departed loved ones who had died before seem to be welcoming them to another realm or in some cases to send them back to life. One of the things about near-death experiences that interest me more as a psychiatrist is deep after effects. People report reliably a consistent pattern of changes in attitudes and values beliefs do not seem to disappear over time. People with near-death experiences tell us that more spiritual after the experience they have and more compassion for others and a greater desire to help others, a greater appreciation for life and a greater sense of meaning or purpose in life. An overwhelming majority of near-death experiences report that they have a stronger belief that we survived the death of the body and like many report that no longer have any fear of death. About half report that they have completely lost interest in material possessions and many report that no longer have any interest in personal prestige or professional competence.

There are different and varied ways of describing consciousness and all are based on interests of different research groups: psychologists, neurophysiologists, computer scientists, philosophers and physicists.

Consciousness can be defined consciousness as the set of subjective, immediate or remote knowledge that each being has about the world and himself. There are three main schools that explain consciousness:

4.1.1 School neuroscientist

Obviously proponents of this school advocate that consciousness arises from an activity neuronal merely more or less complex, and thus resides in the brain. This is the toughest school in their criticism of the others, which sometimes does not hesitate to ridicule because they are considered unscientific and too uncritical. Despite their sharp historical background, this is a real fashion. Its main champion is Francis Crick (Crick & Koch 2003) has been dedicated the last twenty years to neurology. Its basic principle claims that before treating consciousness as something mysterious and spurious, we must investigate under strictly scientific premises. If we declare the brain as an organ of unknown structure and possibly irrelevant, we never get to study and thus meet its root functions regarding consciousness. Crick and Koch (2003) propose a thorough study of neurons and their interactions that would pinpoint scientific models of consciousness, image of what happened with the transmission of genetic information through DNA. To do this, Crick and Koch have focused their studies on the system visual uptake in humans, since it is the best known and whose mapping neuronal is more clarified. If you could come to establish mechanisms of neurological type of consciousness to see, perhaps it would give rise to continue other more complex as self-awareness. The problem is that if "being conscious of itself" is a phenomenon only human, which does not occur in animals as now it seems, the complexity of studies could be practically be an insoluble matter.

M. Edelman (1993), holds that our sense of consciousness arises from what he calls "neural Darwinism" that would not be but maintaining a close fight together of large groups of neurons to configure a representation of the world. However, neither Francis Crick himself has been free of similar criticism, as has been the case Gerald D. Fischbach (2002), a professor at Harvard University and president of the Society Neuroscience, who has made no secret of his view that the proposed Crick regarding a "electrophysiological" lack of awareness sufficient scientific rigor, and that in any case seem too advanced theories taking into considering the current state of maturity of the neurosciences. Tomaso Poggio (Koch et al 1983) from the Massachusetts Institute of Technology, believes Crick is giving undue prominence to the interneuronal excitations which could explain a visual scene, to the detriment of other capabilities of the brain as its plasticity and ability to change network connections may be created by this last system state of human consciousness.

Also within neuroscience school, another researcher, Antonio R. Damasio (1989) of the University of Iowa, disagrees with Crick in the sense that if a comprehensive theory of consciousness should include how to acquire the sense of our self, it should not only take into account the brain, but the whole body. Moreover, he is also convinced of something that otherwise seems obvious: consciousness is also molded along the life of a human being by interactions between him and his physical and social environment, so a model simply neuronal consciousness is doomed to failure if not taken into account other social forms of knowledge and theories. In contrast, Terrence J. Sejnowski, neural network researcher at the Salk Institute, states "All I have heard about arguments against the theories attribute to Crick shows ignorance. " "If researchers would adhere exclusively to Crick research patterns in a short time could really get something. ". Sejnowski also explain the other explanatory tendencies of consciousness, quantum and skeptical, have their place "also has its advantages that the discrepancies are so deep, as well as researchers try to be more original and rigorous in its studies. "

Christof Koch (Koch & Crick 1994), defends the thesis of the neuronal scheme of consciousness based on experiments with animals, mainly with anesthetized cats. These experiments speak of a certain neuronal synchronization in acts of perception and sensitivity, but have the huge problem of the place of "assembly". He briefly explains this point referring to the typical Christmas tree decorated with many little bulbs of different colors. In an ambiguous situation observation, such as a landscape, the neurons would be the bulbs going on and off for no apparent order, chaotically. However, when there is a perception, such as vision of a known face, it happens that a particular group of light bulbs go on in unison (synchronicity) and in a certain specific area (location), it draws attention of the "neuronal group" on the perception received. The problem is that, by one hand, reason is unknown why certain group of neurons coordinate and no other, and on the other hand there seems to be a unique place to decide how to coordinate neurons and even how often they should be.

Other lines criticism of current neuroscience, even within their own breast, they are those held by Walter J. Freeman and Benjamin Libet, on which we will just comment, so as not to tire the reader, that while they accept a first rank role for neuronal system in the phenomenon of consciousness, They believe that this is nothing but a piece of the puzzle of consciousness and "The current wave of enthusiasm about it is out of place" (Libet, Freeman& Sutherland 2000).

4.1.2 Quantum school

The trend to explain consciousness by applying quantum theories has gained popularity in recent years and, although clearly disdained by neuroscientists, more and more researchers direct their steps this way up. Brian D. Josephson (1962) of the University of Cambridge, winner of the 1973 Nobel Prize in Physics for his studies on the quantum effects in superconductors (Josephson effect), proposes a unified field theory, quantum nature, that would explain not only the consciousness and its attributes, but also all the phenomenology observed to date in terms of parapsychological and mystical experiences.

Another character that has stood in defense of a theory quantum of consciousness was the physicist Roger Penrose (1994), famous for its friendly Stephen Hawking differences regarding certain cosmological aspects and a world authority with regard to the theories of relativity and quantum. Penrose (1994) attacks and almost ridicules those who argue that the artificial intelligence of computers can reproduce human attributes, including consciousness.

Penrose, based on the mathematical theorem of Gödel and based on subsequent his elaborations, concludes that no system deterministic, that is, which is based on rules and deductions, they can explain the creative powers of the mind and your judgment. This nullifies the claim of physics classic, computer, neurobiology, etc., to structure themselves into a complex phenomenon of consciousness. Penrose says that only the peculiar characteristics non-deterministic quantum physics could issue an approximate judgment on consciousness, within a theory that involved quantum phenomena, macro physical and conditions of non locality. At this point perhaps it would be interesting clarify that local conditions are not known in quantum physics those capabilities that either has a quantum system, communication instant between parts, ie, without there being time duration between communication of an event from one point to another system.

Albert Einstein called this peculiarity that occurs in the innermost parts of matter, the universe, "that mysterious action at a distance '. If some concepts explained at the time in the course of "Matter and Energy" understand the model proposed by Penrose relates to the concept of Universe "Nonlocal" to quantum levels, in the sense that everything that happens in a corner any of our cosmos is immediately 'felt', "meaning" for any another.

In his last conference Penrose ventures even indicate that probably It is in microtubules, microscopic tubes that form the skeleton of cells, including neurons, where the complicated interactions occur quantum kind that give their "magic" character, "mysterious" from the point of view of science, consciousness.

Hameroff (1994), claims to have found evidence that loss of consciousness by providing anesthesia is due to some inhibition of the flow and movement of electrons within the microtubules. Hameroff argues that certain cellular elements such fluctuations occur quantum-relativistic that "somehow" do emerge consciousness. The major objection to this theory by neuroscientists is that all animals, including elementary, have microtubules in the cells which It seems to imply that they all possess consciousness. Hamenoff argues that such statement would be indefensible, but it is inevitable the observation of "some degree of apparent intelligence" in all animal species. There is still another favorable group to this explanatory theory of consciousness, which is headed by Dr. Ian N. Marshall (Marshall and Zohar 1997) who through empirical testing system claims to have the key to the issue. Marshall and Zohar (1997) showed that conscious thought emerges from quantum effects. Indeed, it has been found that the ability of subjects to carry out work simple while they are connected to an electroencephalogram (EEG).In general, we can say that the Achilles tendon of this group of theories lies in the extreme conditions in which the quantum interactions are observable, for example, the effect of local not manifested in conditions close to absolute zero temperature is absolutely clear that they are not exactly those of our brains. However, supporters of quantum school, as expected, they find answers to these problems and many other objections, in which are indistinguishable from the other schools that obviously do the same to defend their ideas.

4.1.3 School skeptic

We call the skeptical school as the one whose followers hold that science can never interpret and understand consciousness. In general, this would be so because the secret of "being aware" is not based on a simple phenomenological problem, but on the contrary, the great challenge is to explain that part of "the consciousness that is aware of own consciousness". In other words, the great mystery is that we are aware of we have consciousness, and that is irreducible to science.

This way of seeing things has its followers even among members of other schools, as Penrose, Josephson, and in general those researchers outside the disciplines of neuroscience, as physicists, philosophers, etc. Neuroscientists initially despised these eclectic ideas, then try to ridicule, and finally, in the last three or four years, passing discuss it to the undeniable intellectual stature of many of his followers.

Jerry A. Fodor (1992), philosopher, professor of Rutgers University, doubts, indeed, that no theory based on purely materialistic aspects can never explain why humans have a subjective experience of it, and we also realize it. ". The question is how you can have any physical system, like our body, as a conscious state "those who think that science alone can explain means that you do not really understand what is consciousness".

Another philosopher, professor at Duke University, Dr. Flanagan (1954) says all tests so far carried out empirically(ie based on actual experiences under certain control), actually test nothing concrete since in all cases it was people especially trained to do this or that exercise, which was to be measured. Such training distorts the conscious content of the individual, so nothing can be said about consciousness in such a case. In fact, the Dr. Flanagan argues that it is possible to talk about different types of consciousness, already even neuroscientists have so far been able to confirm that the system neuronal perceived, for example, aromas, is different from that responsible for visual perception. Flanagan is an advocate of a broader theory, which he calls "Constructive naturalism '", according to which consciousness would not only be in the man, but also in other animal species and especially primates. The scientifically elucidate these differences you crave task beyond the reach of human beings.

Daniel C. Dennett of Tufts University supports Dr. Flanagan's line, in his book entitled "Consciousness Explained" (Dennet 1993) he proposes, according to the latter, that if something can glimpse regarding consciousness is a triple system that integrates the neural data, psychological and those deduced from human subjective experience. This scheme would accommodate some species of higher level animals. Within the group we might call "skeptic" of this line of thought skeptic, is Colin McGinn, doctor of philosophy and professor of Rutgers University. In his book "The Problem of Consciousness", makes clear his argument that we are not equipped to understand the workings of consciousness, despite its objective naturalness (McGinn 1991). Thus, in the same way that any animal species can not even guess the meaning of a football game, a bet Betting Community, a single lottery ticket, may be that the human species will be off limits certain areas of their existence, including the mind-matter relationship would be found.

McGinn explains that any theory that gird strictly to physics, biology, etc., can never explain the meaning of the consciousness. In any case, these disciplines can resolve problems regarding the specific brain functions, but none of them can justify such brain functions as these are accompanied by subjective experiences. He also argues that just as science possess experimental concepts as space, time, load, field, etc., the theory of consciousness is to start managing others who themselves, help with its definition and understanding, such as the proposed:"concept of information", as it brings together physical and experiential or subjective aspects.

Koch, C., & Crick, F. (1994) proposed some further ideas regarding the Neuronal Basis of Awareness. Neuroscientists in principle are not satisfied with anything that smacks -of subject- subjectivity, and no longer be right when opposed, on the road undertaken by MdGinn and

Chalmers, we could reach undesirable prophecy it is self-fulfilling. Indeed, if we accept that science has nothing to say about consciousness and so we stopped investigate ultimately not

5 NEURO AND BIOENERGY TECHNOLOGIES

Interest in neutechnological devices has grown strongly since the release of 1986's classic text of Michael Hutchison "Megabrain" (1986). Besides neurotechnological machines, the industry has developed a range of related devices including peripherals-tapes, subliminal tapes, hypno-peripheral processing, sound meditation devices and altered state induction software programs for personal computers, etc. It seems like neurotechnology combined with contemporary psychology may seem to be revolutionary but its roots are much older as humanity has made many attempts in the past to involve consciousness with technology. The flashing LEDs in a light and sound machine can be traced back to the old practice of starting a fire in order to induce trance. Sound recordings of brain entrainment undoubtedly have their roots in the ritual singing and shamanic use of wind instruments.

This current era of neurotechnology was born from the experiments made in biofeedback that resulted in the work of some researchers of consciousness as Masters & Houston (1966) that created the AVE (an abbreviation of Audio Visual Environment) that is one of the first electronic brain entrainment devices. The AVE was the first device to use visual and auditory stimuli as a means of altering consciousness. Not much different that many of audiovisual devices of today, the AVE consisted of a wrap-around screen and headphones to bomb the sense in these two stimuli.

These signals generated by AVE devices and detectable by electroencephalogram (EEG) tests at specific frequency ranges; seem to vary from one state of consciousness to another for their users. The products of neurotechnology appear to make possible to "tune" the brain to these different alternating frequencies, and in turn achieve new psychological states. When you "tune" into these frequencies at a slower pace, our thought processes and central nervous system are altered, causing the brain to fall into a deep level of activity. In addition, when one is tuned to a frequency faster and a higher range, the wave activity can have a highly stimulating effect on the brain.

Altered states of consciousness have been used as a therapy for modern psychological paradigms as transpersonal psychology (Valverde, 2015). Although there are several techniques such as holotropic breathing used to induce these altered states of consciousness, terotechnology offers an easier to control alternative to the transpersonal therapies.

The area called Neuro includes all types of technology that any it contributes to the advancement of knowledge related to the brain. Despite that neurotechnology is relatively old, its most significant advances occurred in the last 30 years. The 90s were even proclaimed by the US government American as the "Decade of the Brain". Much of the responsibility for this development exacerbated is due to the advent and popularization of brain imaging techniques, including MRI, CT, PET (Positron Emission Tomography), among others. With technological advances, some of these techniques at first provided only structural information from the brain but now it allows us to study the brain dynamics.

Experimentation and published research on the altering of brain activity has been going on for decades. A vast amount of applicable data has been collected by researchers and scientists concerning the manipulation and influencing of brain waves and states of consciousness. One effect discovered by German scientists in the 1930's is called the Ganzfield effect (Hyman 1985). It is an effect in which the eyes are exposed to a totally uniform visual field, with no edges, color changes, or movement. This effect can be achieved by looking up at the sky on a clear, cloudless day or with very specially constructed illuminated glasses. It affects the circuitry in the brain, which normally is constantly scanning the visual field for edges and movement. The eye responds primarily to changes, and with visual sensations shut off and the brain finding absolutely nothing, the subject cannot tell if his eyes are open or closed. This is an example of an altered state of consciousness resulting from the shift in brain waves to the theta state.

The effects of flickering light stimulation have been known for centuries. Ancient shamans used flickering lights to achieve a trance state, a key component in the development of the spiritual practices of a wide variety of ancient cultures. In the literature of the common era, Ptolemy noted around 200 A.D. that the flickering of sunlight seen through the spokes of a spinning wheel could cause patterns and colors to appear to the observer, producing a feeling of light headedness and euphoria. The onset of modern times, (1940's and 50's) brought the research of neuroscientist W. Gray Walter (Walter and Walter 1949). Walter used a strobe light to create flickering light stimulation, noting that the brain wave pattern of the cortex—just the area associated with vision—was changed. Ever since the adrenaline rush brought about by dancing around the tribal fire, the intent of scientists, researchers, and ancient religious practitioners has been to create relaxation and altered states by affecting the vibrational frequency of the brain waves. The use of technology is a modern day attempt to create desired changes in the brain's functioning, bringing about states of well-being and health.

Robert Monroe, one of the most prominent people in neurotechnology in particular in the field of sound with neurological efficiency, devised the method HEMI-SYNC: (Synchronization of the cerebral hemispheres by means of sounds). Like a glass resonates when a pure tone is emitted, the brain resonates when receives certain frequencies of waves, being synchronized with these, (similar to the previous with the flashes of light) effect is known as FFR Frequency Following Response (Response a frequency tracking).

Machines created based on FFR began to be popular, the typical machine based on the principle of Monroe using stereo headphones that are used are separately sent sound to each ear signals, for example 2 signs of 300 and 304 Hz. In one ear will be heard only 300 Hz signal and the other only 304, but since the sounds are combined in the brain, this will hear a third signal of 4 Hz which is the difference between the two sound impulses.

This third sign is not an audible sound, but an electrical signal that can only be created by the cerebral hemispheres acting in unison, and may go unnoticed, producing as a result, the two hemispheres are focused simultaneously on the same state of consciousness, increasing and brain power.

The Monroe Institute was created based on the principles of Monroe. This institute was founded by Robert A. Monroe, who began studying the effects of sound on accelerated learning in the 50's.The work is the result of thousands of hours of laboratory sessions.

Participants listen volunteer's specific combinations and sequences of sound patterns and simultaneously are reporting their experiences; while electronic instruments measure the effects on the activity of their brain waves.

W. Grey Walter (1964) conducting experiments in which strobe devices used to send rhythmic flashes of light in the eyes, watched amazed that the flickering light could alter the activity of the entire cerebral cortex, rather than exclusively associated areas with vision. The subjective experiences of those who received the flashes were even more curious reported seeing "lights like comets, ultra fantastic colors, color mental". Moreover, Walter discovered that certain visual stimuli of a certain frequency could cause the brain to respond quickly adjusted to the same frequency of the stimulus it received.

In 1982 a psychiatrist Denis Gorges created a machine called brain wave synchronized energized, patented and sold by his company, Synchro-Tech Inc which applies the latest technology in the first device designed optical-acoustic stimulation.

This is able to promote as its inventor claimed, "an increase in the capabilities and functions of the human mind" unleashed a wave of unprecedented research and experimentation, catalyzed by M. Hutchinson (1986) through his famous book Megabrain that is being developed in and outside universities around the world.

Studies on this type of equipment is underway in North American universities, in areas including education and training, substance abuse, gerontology, pharmacology and physiology as well as pain reduction. In Europe, at the University of Vienna, it has created a special subject for further study of these devices. The neurological investigations in the last 15 years have shown that our cerebral hemispheres work independently. Each hemisphere collects the same information that its sensors (eyes, ears, taste, touch, smell) and processed differently.

Our scheme of thought is influenced by the predominance of left hemisphere, logical and analytical thinking. Right hemisphere, creative thinking, visualization, and which carries out a holistic synthesis.

On the back of the brain is where the intuitive thought (Subconscious). Obviously use one of the two hemispheres or either of the Brain regions severely limits our abilities. The optimal functioning of our brain, is given by the synchronization of the two hemispheres: By left-right and front-back alternating stimulation occurs almost simultaneously, an issue of brain wave amplitude and frequency identical. It is that how you might think with our whole brain.

Neurotechnology are typically used for process brainwave entrainment to tune brainwaves any range brainwave. With these machines, you may experience theta waves, alpha, delta or including combinations of ranges using frequencies layered that mix several ranges brainwave in a brainwave pattern synergistic as pattern brainwave.

Hemi-Sync is a patented process sound pulses to create the same electric waves in frequency and amplitude simultaneously in both hemispheres of the brain for specific purposes.

Our brains have a hemisphere left and right. The left brain is linear, logical, practical, and time - oriented. The right hemisphere seems to be much more nonlinear, abstract, creative, holistic, and not logic.

We tend to use one hemisphere at a time, or better said, will favor hemispheres particular in terms of what we're doing. An accountant probably uses less of his right hemisphere that an artist would during the course of their workday.

If the person is doing math, the staff would be using more of your left hand. If the person is going to paint a picture, the person would have the activity hemispheric more right. Both hemispheres are constantly interacting and both can be in use simultaneously.

By merging both hemispheres, a person is allowed to work together and with this the person can increase his or her ability mental and improve his or her performance cognitive in general. Basically it's like having a computer processor faster able to work at faster speeds. The greater integration creates better performance.

The mind machines that help the brain synchronizes naturally to balance hemispheric activity and adjusts the activity of brain waves to match the carrier frequencies embedded brainwaves. This hemispheric coherence induced audio produces an optimal state of comprehensive synergy of the entire brain.

These hemispheres are connected by the corpus callosum. It serves as a conduit or a bridge between both sides. This bridge literally be exercised and strengthened until it is physically larger and more capable of transmitting data, ideas and feedback between the hemispheres.

The advance occurs when you use this principle to synchronize our brain wave frequencies chosen with the help of specific mental machines. We can do this easily by using technology binaural beat audio and ringtones monophonic. This phenomenon is called trailing phenomenon.

The scientific principle of drag can be used to resonate, or synchronize your brain to tune specific frequencies. The drive is quite simple. If you have two tuning forks of the same pitch, if you strike one and hold it near the other, which both resonate at the same frequency. Given this fact and the fact that our brains function as a resonating chamber. Oscillating pulses and patterns of neural excitation ripple through our brains like endless waves on a pond of subtle dynamic electric field.

The creation process frequency binaural beat - brainwave entrainment is very simple to explain. First, you create two audio tones two frequencies slightly different (ie 100 Hz and 107 Hz). When playing a tone in the ear left and the other tone in his right ear, the brain reacts to the stimulus audio by resonance brain activity to the difference between these two tones. His brain waves adjust to match this frequency differential. For example, if you play a frequency of 100 Hz in the ear and left 107 Hz in the right ear , your brain tune the 7 Hz differential (107 Hz - 100 Hz = 7 Hz). This frequency tracking response makes your brain to resonate at 7 Hz frequency falls within the range of brainwaves theta . This guide psychoacoustics state of the art can create phenomenal results.

The electrical activity of neurons is changing patterns of electrical potential through all cell membranes. Each cell generates in the membrane potential can be detected in about one micrometer of the cell but large sets of brain cells that can work in harmony to generate potential of few micro volts, that can be detected through the skull and scalp using electrodes. The variation of the electrical potential around the scalp is the basis of operation of the electroencephalogram. This method of recording the electrical activity of the brain, digitized maps produces the electrical activity of brain.

The state in which the two hemispheres are synchronized, that is, showing similar activity in both hemispheres, sometimes known as "whole brain" naturally occurring randomly for short periods. Stimulation through the differential tone makes it possible to sustain this state for longer periods.

EEG based on data collected in the laboratory of Gateway Institute, it is clear that after a brief period of use of frequencies of Hemi-Sync, it can be demonstrated the benefits of the technology without the use of external sources.

The Hemi-Sync process has already been tested and implemented in many ways for:

> Sleep better
> Reduce tension or stress
> Control pain

Accelerate learning

Study and concentration

Enhance creativity

Find solutions to problems

Light and Sound Machines have the purpose to create a relaxed state through the use of sound and light stimulation. It was discovered in the 1930's that repetitive light stimulation (strobing) caused brain waves to follow and pulse at the same frequency (Walter & Walter 1949). This discovery had profound implications for the human condition, giving rise to an entire industry that explores the possibilities in using outside stimulation to provide a functioning aid to the workings of the brain. The devices on the market today are set at predetermined frequencies of 1 to 40 Hz. The user is then free to experiment on just how these vibrational frequencies influence their affect on the mind's eye. Most users report patterns of colors that pulse in synch with the frequency of the beats. The patterns are often quite psychedelic and geometrical.

Light and sound machines are often used for meditation assistance, or to aid the learning process. The technologies are referred to as audio-visual stimulation systems, light and sound machines, or mind machines.

Varying patterns of light and sound are used to create different states of awareness, attention, relaxation, creativity, and receptivity to information. Most systems are controlled by a microprocessor. The sound and music sometimes comes on a CD, with the frequency patterns embedded in the music.

A typical system is made up of three components: a light/sound synthesizer, stereo headphones, and stimulation glasses with pulsating lights. A carefully composed sequence of rhythmic light and sound patterns direct the stimulation synthesis, creating light and or sound sessions for the desired effect. Some systems connect to your music CD player, or have one built in. The process employs a technology called an audio strobe system, which creates inaudible, high frequency beats that drive a light/sound system. Other systems have computer interface capabilities that allow the user to vary the tone, frequency, pitch and other parameters to tailor the program to individual needs. Some metaphysical technology companies have downloadable programs that allow for even more growth, variety, and individual tailoring. A few make a point of offering customer service where the company provides you with specifics that are designed to individual needs (Dumit 1995).

The light and sound systems are designed to blank out the visual and auditory stimulation that the mind generally receives, and replaces it with very simple, rhythmic stimulation. Some light systems come with glasses that block certain colors so that the brain focuses on others to create the desired stimulation. The basic rhythms of light and sound cause the mental processes to change in response. Thus, the rate of brain waves adjusts to the desired vibrational rate.

Another technology designed to create specific body changes is biofeedback. Biofeedback is a non- invasive, painless, non-pharmaceutical method of teaching you to become aware of your body processes and help you learn to change them. Electrodes are placed on the skin and feed information into a small box attached to your computer via a serial port. The equipment feeds information back to you about your body. The feedback is usually auditory, a soft beep, and, by concentrating on this sound, the user can focus on specific bodily processes and learn to relax in very specific ways. The process is designed to provide the individual with an ongoing progress report as to how their body is responding to their own internal controls. In this way, one can learn to change heart rate, skin temperature, muscle tension, and certain brain wave patterns (Turnbull & Ritvo 1992). Biofeedback information is used to teach the patient what are the brain regions responsible for a certain behavior and thus be able to learn to control these areas and monitor the progress of the training.

The difference between the two technologies is that light and sound machines provide direct stimulation that affects the rate of brain waves, whereas biofeedback places the impetus on the individual to control their body's processes themselves. The goal of biofeedback is to train oneself to respond to intrinsic cues, which contrasts to the light sound stimulation machine's purpose of matching one's brain frequencies to outside stimulation.

Neurothecnology is for healing, and affecting one's mental and physical state in a positive way. However, in order to best apply the technologies to one's personal circumstances, diagnostic processes may be needed in regards to one's individual circumstances. One such diagnostic tool is called Kirlian Photography.

This process was discovered in 1939 when Semyon Kirlian (1973) found that if a living object on a photographic plate is subjected to a high-voltage electric field, an image is created on that plate. The image looks like a colored halo or coronal discharge. This process captures what the spiritual community refers to as the "aura," and the scientific community calls the "Meissner" field. This field is the area of gas ionization that is produced when electricity enters a living object. The image of the moisture is transferred from the subject to the emulsion surface of the photographic film. The field is visible to the photographic process because of the exchange of photons, or light particles, which takes place as our cells communicate through the chemical process with each other and with the surrounding environment. The light particles the body produces carry information in the same way that fiber optics carry information from one point to the next.

The color and densities that appear on the photo will vary due to the variances of energy output as a result of the different quantities and balances of acids and alkalis the body gives off as it attempts to optimize its chemical processes. The appearance of this gas ionization process on film varies in ways that are consistent. Thus, diagnosis of one's chemical balance can be inferred from the colors and densities that appear on the film.

Therefore, a good interpreter can detect specific energy imbalances from the color variances in the field surrounding the image of the body. Pulsed electromagnetic fields—which induce measurable electric fields—have been demonstrated to be effective for treating slow-healing fractures and have shown promise for a number of other conditions. This has opened up a huge effort to market a wide variety of magnetic products to assist with the healing process. A whole range of products is available for what is referred to as "magnetic therapy." Small spot magnets can be used at the point of pain on the body. There are wraps, belts and pads, which are held on the body in some fashion. In addition, there are mattress pads that claim to balance the body's energies while one sleeps. Various types of jewelry incorporate magnets into their design, including a wide variety of bracelets and necklaces. Many magnetic necklaces, bracelets, and earrings are formed from silver and gold-rich magnetic alloys and are promoted as both fashionable and therapeutic. One catalog claims that magnetic earrings "stimulate nerve endings that are associated with head and neck pain," and magnetic bracelets "act upon the body's energy field" and "correct energy imbalances brought by electro-magnetic contamination or atmospheric changes" (Vallbona, C., & Richards 1999).

The broadest explanation of the use of magnets in treating physical ailments was presented by Dr. Kyochi Nakagawa of Japan (Nakagawa, 1976). Dr. Nakagawa claims that many of our modern ills result from "Magnetic Field Deficiency Syndrome." The earth's magnetic field is known to have decreased about 6 percent since 1830, and indirect evidence suggests that it may have decreased as much as 30 percent over the last millennium. He argues that magnetic therapy simply provides some of the magnetic field that the earth has lost. Magnetic therapy is also prominent in the treatment of thoroughbred racehorses. The incentive to try "alternative medicine" to supplement mainstream veterinary treatment is considerable; an injured racehorse represents potential loss of a substantial investment. Magnetic pads, magnetic blankets, magnetic hoof pads, and more are used for a variety of leg problems. All get ringing endorsements from many horse trainers, and even some veterinarians. One marketer of magnetic products for humans reports that he first became convinced of their effectiveness when he used them on his llama. Some argue that the placebo effect might be taking place in these circumstances, but the psychological aspects of the healing treatments are obviously not happening with the animal. However, magnet therapy enthusiasts forget that it may influence the human who is interpreting the effect of magnetic therapy on the animal.

Clinical studies report between 75-85% of patients experience relief using magnetic therapy. Whether or not the placebo effect is taking place is always part of the debate. One point of contention in the argument is that the magnetic back braces used by many senior golfers may help ease their back pains through providing mechanical support, through localized warming, and through constant reminder to the aging athletes that they are no longer young and should not overexert their muscles. All these effects are helpful with or without magnets. One British study of pulsed-field bone-growth stimulators, which were approved decades ago by the FDA, found that they were equally successful when the devices were not activated (Barker 1984). It concluded that their effectiveness resulted from the enforced inactivity associated with their use rather than from the pulsed magnetic fields. The logical conclusion is that with circulation increased, the body's natural healing mechanism can work more efficiently. Some doctors believe that the magnetic field affects the iron component of blood, hemoglobin, thereby increasing circulation to the area where magnets are applied. Another theory is that the magnetic field energizes and oxygenates the white corpuscles in the blood stream, and these white corpuscles are nature's healing agents. In this regard, the charged ions increase the blood flow, which provides increased oxygenation to the blood. This increased oxygenation is the prevalent factor to enhancing your body's natural healing powers, and the results are less pain, decreased inflammation, and the possibility of increased energy levels. Other specialists believe that, since nervous signals travel via tiny electromagnetic charges, an electromagnetic field may actually tune out "false" signals. The actual pain relief mechanism of magnetic energy may never be truly understood, since it operates within the tissues at such a microscopic level. What is clear, however, is that belief is necessary for effective healing.

Studies of magnetic therapy practices and results are more common in Eastern Europe and Asia where there is less access to pharmaceutical drugs. There is evidence that magnetic therapy does improve circulation and oxygenation in the bloodstream, and thus it would serve the purpose of facilitating healing for areas of the body recovering from injury or weakened by age or disease.

One property of magnetic energy is that it has no measurable mass. This is also true for another kind of energy relevant to this course, Orgone energy. Orgone energy was objectified and scientifically demonstrated by Dr. Wilheim Reich (1951). Reich built on Freud's early work, which strongly suggested that emotions and sexuality were expressions of a tangible energetic "something." It was Reich who provided the clearest evidence that the Freudian libido was a real energy, discharged during emotional expression and sexual orgasm. Reich discovered that this same discharging energy was present in a single cell, which apparently caused him to question whether that energy could be collected and harnessed.

In 1940 Reich discovered a way to accumulate that energy, and labelled it Orgone Energy. His work was not taken seriously at the time. There was no social context providing an atmosphere of acceptance for the idea of energy from repressed sexuality. However, a simple Internet search reveals 23 published articles on the treatment of a wide variety of ailments, from cancer to rheumatoid arthritis, by the application of Orgone energy. However, despite being used to treat specific ailments since at least 1971, there are numerous qualifiers in the dialogue on its applications. The point is frequently presented that science and society will have to go through some significant changes before its use will be accepted by the general public. Orgone energy is the primordial life energy, the fundamental creative force long known to people who are in touch with nature. Perceived as "chi" by acupuncturists, Orgone energy is accumulated naturally in biological organisms by food, water, breathing, and through the skin. In 1940 Reich discovered a way to collect or "accumulate" Orgone energy from the atmosphere. Reich's research revealed that energy is attracted to and absorbed by organic substances, but is attracted to and instantly re-radiated by metallic substances. He developed a special enclosure resulting from his experiments with the Faraday cage, a device built in 1836 by Michael Faraday to exclude energy. Reich's accumulator was made from layers of organic materials which absorb and hold the energy, and inorganic materials which attract and then rapidly repel Orgone energy. He found that the inside of a metal-lined enclosure that is layered with organic and inorganic materials contained a higher concentration of Orgone energy which was detectable by a higher temperature reading inside the accumulator.

Reich's device allowed him to isolate Orgone radiation from its environment, and then accumulate the energy within his device. Orgone radiation is not foreign to the body. The Orgone energy in the atmosphere is just another form of the same Orgone energy within the human body. The accumulator concentrates that energy and helps the body to help itself. Reich discovered that the energy moves in the direction of a magnetic field. Thus, when a person with their own energy field comes into contact with an accumulating device, the two fields make contact and excite each other. This creates a stronger excitation called "lumination." This strong, orgonotic charge has been found to be very beneficial to living systems. It may help to strengthen the immune system, improve circulation, and raise one's energy level. Investigations have shown that contact with energy from the accumulator stimulates the body's parasympathetic nervous system, inducing relaxation and a feeling of expansion.

Modern research has found ways to manipulate and focus the energy, thus resulting in applications to the medical field. The motion is a pulsating expansion and contraction, and a flow normally along a curved path. Inside an accumulator, the energy is emitted as a spinning, pulsating wave. A wide range of devices, and even blankets, are available on the market that makes use of this technology.

Another use of technology that has its roots in Asia is the adaptation of modern technology to the ancient art of acupuncture. Electro-acupuncture is the application of a pulsating electrical current to acupuncture needles as a means of stimulating the acupoints. The acupoints connect the individual's body to the universal, cosmic energy that connects us all. The stimulation of these points has been an ancient healing tradition that reaches back through the millennia. In 1934, according to Subhuti Dharmananda (2003), the Director of the Institute for Traditional Medicine in Portland, Oregon, Chinese healers added electronic stimulus to the process as an extension of hand manipulation of the acupuncture needles. The process is described, though only briefly, in most comprehensive texts of acupuncture. The procedure for electro-acupuncture is to insert the acupuncture needle as would normally be done, attain the qi reaction by hand manipulation, and then attach an electrode to the needle to provide continued stimulation. The idea was to help ensure that the practitioner, who may otherwise pause due to fatigue, gives the patient the amount of stimulation needed. It was also thought that Electro-acupuncture could produce a stronger stimulation without causing tissue damage associated with the twirling and thrusting of a needle. The stronger stimulus provided by the electronic current is better for treating more difficult cases of paralysis. Positive results have also been seen from treating patients with the chronic pain and spasms associated with neurological diseases.

The application of modern day electronic technology to an ancient healing art was developed to reduce total treatment time by providing a specific continued stimulus. The assumption was that it would be easier to control the frequency of the stimulus and the amount of stimulus than with hand manipulation of the needles. An added benefit was that it would free the practitioner to attend to other patients. However, the subsequent inattention to the patient receiving the electro-acupuncture limits the opportunities to respond to changes that are taking place during treatment. In addition, the literature cautions against using the technique on patients with serious cardiac diseases. It is generally recommended to avoid placing electrodes near the heart, as the heart can respond adversely to electrical impulses.

Another use of electrical impulses in the healing arts is Cranial-Electro Stimulation machines (Gilula & Kirsch 2005). The use of these machines involves placing sensors under the ears to send a minute electrical impulse. A slight tingling sensation on the skin is felt. The stimulation has the affect of tricking the brain into releasing the body chemical serotonin. Serotonin is a neurotransmitter that plays an important role in the biochemistry of depression, migraine headaches, bipolar disorders, and anxiety. There is also evidence that its presence or absence in the body influences sexuality and the appetite. The serotonin released can help relieve anxiety, tension, and the feeling of being "on edge."

There are portable versions of this machine, powered by a 9 Volt battery, that are small enough to put in a pocket. The sensors are placed on the neck below the ears, and can be used while walking or using the computer. They are designed to be used once a day in response to feelings of anxiety. It generally provides adjustments to the body chemistry and the ensuing relief from anxiety within 5 to 10 minutes.

Another device in the category of neuro technology facilitates lucid dreaming. Lucid dreaming is being aware of the fact that you are dreaming while you are doing so. This allows the dreamer to gain insight into the purpose and symbolic meanings behind the dream, thus facilitating their analysis and the ensuing psychological healing process.

The technology for this process was developed by Stephen LaBerge, Ph.D. (1990), during twenty years of research at Stanford University. Dr. LaBerge developed a device called the NovaDreamer, a mask-like device that detects the time that one is in REM sleep, the state of rapid eye movement that one experiences when dreaming.

The device gives cues, such as flashing lights or specific sounds, that remind one to realize he or she is dreaming. In other words, one "wakes up" while still in the dream. The purpose is to give the dreamer detachment from the dreaming process during the course of the dream. This provides the impetus for the analytical process of dream interpretation to take place during the course of the dream. Users have reported that having lucid dreams have allowed them to make spontaneous choices that enable them to learn from, and even guide, the dream to constructive ends. Anecdotal evidence of these choices includes suddenly deciding to fly away from situations, empowering the dreamer to feel safer and have more free choice. There are also reports of incidents of making an old friend appear and talking to that person about consequential matters or resolving conflicts. Dreamers report being able to feel textures, hear sounds and smell odors in the dreamscape. The users reported that these sensations were as real as any in the physical world, and they were able to make conscious decisions and choices.

The entire process of lucid dreaming gives the spiritual sojourner tools and techniques, developed deep within the psyche, to explore and use their dreams for self-discovery, creativity, fantasy fulfillment, emotional healing, and profound spiritual insights.

Full Spectrum Lighting, another simple use of technology is the use of full spectrum lighting (Veitch et al. 1991). Full spectrum lighting is a process that uses specifically designed lighting fixtures that mimic the full spectrum of colors provided by our sun's rays. The purpose is to treat Seasonal Affective Disorder, a psychological malady that is brought on by the deprivation of light. This is generally a result of the changing of summer, with its bright sunshiny days, to the fall season, with its shorter and darker day-time periods. The onset of this change can result in depression and lack of energy (Veitch et al. 1991). The use of full spectrum lighting fixtures allows for the same kind of light energy that the sun offers.

The ultimate achievement so far in modern day development of neurotechnology is the Hemi-Sync device, or Holistic Sensory Response Audio.

In 1956, Robert A. Monroe founded the Monroe Institute to study the effects of various sound patterns on human consciousness, including the feasibility of learning during sleep. Monroe was a highly successful businessman, the owner of several radio and television stations. In 1958 he began to have out of body experiences in which, while relaxing or attempting to fall asleep at night, he found himself hovering above his body. At the time, he thought he was having some kind of health problem, but was reticent to discuss it openly for fear it would interfere with his dealings with a very conservative business community. Monroe soon began to quietly but thoroughly examine the circumstances of his experiences. The discoveries made in exploring this phenomenon were validated later by the work of Dr. Melvin Morse and others in their work exploring Near Death Experiences. But as the work continued, and technology was developed to stimulate out of body experiences, the information gathered was to have a profound affect on the whole of our collective perception of the nature of reality (Monroe 1977).

In 1976 Monroe was issued an original patent for technology that altered brain states through sound. Hemi-Sync produced binaural beats, which are a series of barely audible tones which play in the background of soft music, barely noticeable beneath the music. The beats are at one rhythm in one ear, and a different rhythm in the other, producing a combination that takes the brain to a specific output rate of brain waves. The two hemispheres of the brain act in unison to "hear" a third signal—the difference between the two tones. The tones counter balance each other in a way that forms an anchor, or a signal which helps you quickly and easily re-enter the trance state when your unconscious mind begins to hear them in the background. The tones begin by synchronizing with the brain waves associated with the normal awakened state of the mind, and then slowly changing them to assist the listener in reaching the "alpha" level. The signal can only be perceived within the brain by both brain hemispheres working together. The result is a focused, whole-brain state known as hemispheric synchronization, or "Hemi-Sync." This relaxed level was the exact state of mind, it was discovered, to facilitate not only healing, but out of body experiences (Monroe 1977).

The process has been subjected to decades of research and thousands of lab sessions, and has been refined with over 40 years of research and development. Ongoing experimentation, data collection, and analysis are conducted at the institute's laboratory facilities to demonstrate the correlations between subjective experiential reports and objective electronic measurements. Healing benefits have been extensively documented, ranging from enhanced alertness to relief from insomnia and the myriad of circumstances that fall between.

There are also testimonials regarding profound healing experiences. One woman with an undiagnosed brain injury suffered in a car accident reports using Hemi-Sync to reverse a very frustrating condition. After being released from the hospital the same day of an accident that totaled her car, she noticed a severe change in her ability to concentrate. After many months of forgetting how to cook, getting lost, and "losing her train of thought" in a middle of a conversation, she began using Hemi-Sync to produce a state of relaxation. Within weeks, she became aware of regaining her sense of humor. From there she noticed a gradual regaining of her ability to concentrate (Monroe 1977).

The device, which looks like a bulky set of headphones, is designed to move one's consciousness past the limitation of being within a body. The spectacular success this device has had in facilitating astral travel has been documented in Omni magazine, amongst others, and experienced by thousands. The most profound research for the Hemi-Sync device was done by the founder of the Monroe Institute himself. Monroe said, "The greatest illusion is that mankind has limitations." This implies that we have limitless possibilities.

The work of the Monroe Institute, with its profound breakthroughs in the use technology for spiritual exploration, has the potential of enabling vital specifics in the evolution of humankind. Despite Monroe's death in 1995, his institute continues to attract and change forever the visitors to its campus as well as the thousands who have visited the virtual reality of its web site.

The implications of Monroe's work developing the Hemi-Sync device take on a whole new meaning when viewed as to the potential for learning from other times, places, and species. The development of this technology, a technology that enables astral travel and contact with guides and other beings with purposes that are highly relevant to both personal and planetary evolution, brings focus and significance to the whole idea of pursuing the potential ramifications of neuro technologies.

Royal Rife is the inventor of the Rife Machine (Hess 1996), it is said that "one day Royal Rife will be acknowledged as one of history's greatest medical geniuses." In 1902 he discovered the human cancer virus with a microscope that he had to invent himself, since no microscope at the time was powerful enough to detect it. Through endless and exhaustive experimentation and laboratory testing, he discovered that this virus could be killed using certain frequencies that were emitted from a prototype machine he created. He was then successful at destroying cancer in patients who sat within 10 feet of the machine. His legend built from there, and by 1931 he was the most honored among the medical society for his track record of being able to cure cancer almost 100% of the time. The cure for cancer has been found long ago! However, by 1939 these same doctors and scientists who so highly praised him denied that they had ever met Royal Rife. A violent attack was set upon Rife by unseen forces, but they could most likely be identified as the drug companies (through the American Medical Association) who had the most to lose from the Rife Machine, and a miffed Morris Fishbein who tried to buy the rights to Rife's work unsuccessfully. After that, people who worked with him were harmed and poisoned, Rife's papers were mysteriously lost, and his labs were destroyed by arson.

He was shortly indicted and taken to trial for "fraudulent medical practices," and for the most part, his life was undone. He became a ruined man. All the doctors who were associated with him lost their medical licenses if they did not stop using Rife's invention, even though they were having incredible success at curing people of cancer. Rife became an alcoholic in order to deal with the terrible turn of events his life has taken. Rife himself was later killed in 1971 by a very lethal dose of valium at a hospital which some do not believe was an accident.

For a long time, Rife's work lay dormant. However, it has been recently re-visited. Others are picking up where he left off. It has been discovered that far more than just the cancer virus can be killed with the Rife technology. Any virus can be killed. It just has to be identified and the right frequency that kills it must be found.

Others have begun to catalog this list. Work on cataloging viruses and the frequencies that kill them has begun again. There are now thousands of viruses that can be addressed using the Rife Machine. It is not that expensive, either, compared to what the medical costs of drugs, doctor's visits, surgeries and hospitalizations would cost. A Rife Machice could be purchased for anywhere between $500 and $5000, depending on how many bells and whistles you want it to come with. If you do an internet search on Rife Machine, there will be plenty of results to peruse, but the only place you will be able to purchase one is from overseas. It is still illegal to use them in the United States. Almost all other countries in the world allow the use of the Rife Machine, at least for personal purposes or experimental purposes, if not for the public. This does seem quite strange.

A Rife Machine works by generating a field of frequency, or frequencies, and anything within that field that it has been discovered to kill will be killed. It works one of two ways. Some of the less expensive machines entail two metal, or glass, cylinders that you grasp in your hands while frequencies pulse through them, and thus into your body. Some of the more expensive machines entail detailed digital readouts and the ability to diagnose your condition by finding out which viruses are present. You also do not have to hold onto cylinders and you are able to go about your business in the room as the machine generates the frequency field that addresses the virus (Hess 1996).

Kilner (1965) proposed a device that is a way for the naked eye to see the first three layers of the human aura. The device consists of two plates of glass with a dycanin dyed liquid between them. Dicyanin is a coal tar dye. Some say that a subject's aura can be seen through the screen, while others say that one must look at a bright light first, and then look at the subject through the screen.

There have been several doctors in the past years who have used this technology for discerning illnesses in their patients and diagnosing their condition.

6 BIOFEEDBACK

Biofeedback literally means, "responding to life". The work of Miller opened the doors to research processes in biofeedback. Miller (1978) constructed an apparatus equipped with sensitive electrodes connected to a monitor on which the patient could see how the skin temperature behaved, teaching patients to relax and concentrate to get it to relax the smooth vasculature of vessels peripheral blood in order to descend skin temperature.

Biofeedback is the process by which a person learns to influence involuntary body processes to receive physiological data from an electronic device that continuously monitors certain physiological parameters. It is a way of measuring the response to the physical, emotional, mental and spiritual stresses of life. Bodies under high stress are more prone to physical discomfort and even illness. The biofeedback response occurs when the body receives new information about their status (*e.g.,* get 'feedback') and make healthy adjustments to reduce stress and tension. The result is a reduction of the nervous activity and increased vitality. Users of the feedback report a greater sense of well-being and joy.

Biofeedback neurotechnology instruments measure muscle activity, skin temperature, electro-dermal activity (sweat gland activity), respiration, heart rate, heart rate variability, blood pressure, brain electrical activity and blood flow. These technologies are able to capture analog electrical signals from the body and translate those signals into meaningful information through complex algorithmic software that a technician can then decipher. Research shows that biofeedback, alone and in combination with other therapies is effective for treating a variety of medical and psychological disorders. Biofeedback is currently used by doctors, nurses, psychologists, counselors, physical therapists, occupational therapists and other professionals. Biofeedback is also used by computer scientists in order to build human computer interactions (Valverde, 2011).

Studies by Jonas and Levin (1999) show that biofeedback is ideal for patients that are looking for therapies that are softer, less toxic and less invasive. The biofeedback technique is applied in the field of psychology for treating phobias, neurosis, anxiety, depression and insomnia.

Biofeedback is the art of listening to the inner tracks given to us from our own body; by being aware of ourselves we can get to get the real possibility of self-control. The changes that occur in our mind-body can be measured. The scientific basis of the Biofeedback is that humans are capable of observe their body facts that are normally not aware off such as the presence of alpha waves in a graph or screen.

Biofeedback is a particular kind of feedback, feedback from different parts of our body, the brain, the heart, the circulatory system, the different muscle groups, and so on.

The feedback is obtained by hooking the patient up with equipment that can amplify one or a number of his body signals and translate them into readily observable signals. These signals may be metric, light displays, and/or audio tones.

Once a patient can "see" these signals he has the information he needs to begin controlling them. Seeing the signals provides a positive reinforcement so he feels he can change the signals.

Biofeedback is a preferred mode of treatment because it is easier to use then other modalities, you just need the instrument and a patient, no bulky equipment. Safer than chemotherapy (with no side effects), it has been the case when medication is used in conjunction with Biofeedback as in migraine headaches, or epilepsy; however, oftentimes due to the physiological change from Biofeedback, medication has often been lowered or eliminated. It is more efficient because it is a self control model and the responsibility for one's health goes to the patient. It is assumed that the patient does want to get rid of his condition unless of course there is a secondary gain (explain) then psychotherapy may be used. The primary goal of biofeedback is to promote self control of physiological processes.

There are three requirements for successful biofeedback training:

The symptom to be controlled must be constantly monitored with sufficient sensitivity to detect moment by moment changes. e.g. especially important in muscular re education and thermal training to enhance reinforcement.

Changes in the physiological measure must be fed back immediately to the patient attempting to control the symptom, again positive reinforcement.

The patient must be motivated to learn the process.

It sometimes is wise to try another mode if a patient becomes stuck at a certain point and is unable to progress. For example if a patient is working on thermal training and becomes stuck, try using EMG to allow the patient to feel progress from muscular tension, then go back to thermal. This procedure sometimes "unsticks" a sticky situation. This situation is different from the patient who does not care about treatment and does not try during the sessions and does no "homework". Oftentimes a. patient will not be appropriate for Biofeedback training and will have to be terminated from treatment.

By Biofeedback, humans can control their states of consciousness, like the mystic who follows the methodology or the yogi zen, only in the human case it would be a mystic manipulated control. On another level, and in their definition, this is a cognitive learning process that is directed inwardly.

The instruments most commonly used in the practice of the Biofeedback are (Sonty 2003):

- EEG, which measures the brain's electrical activity.

- ESR, which is able to measure electrical resistance of the skin.

- EMG, which detects the electrical impulses in muscle tension.

- Thermistor, a temperature measuring skin area.

- Mental mirror, measures the electrical resistance of the skin and brain waves,

We are thus trying to read the mind and the body directly with the use of Biofeedback. We must point out that, phylogenetically, the brain can be distinguished with the following components (Eurler 1986):

- Brain Stem, also identifiable in the reptilian period.

- Limbic system, cerebellum more added to the brainstem.

- Cortex.

- Branch in hemispheres and connections.

Regarding the latter aspect, we would consider that usually the left hemisphere is considered analytical, logical, showing activity in times sequential. The right hemisphere, by contrast, is considered the artist, and therefore creative and abstract. Many people use only the left hemisphere. However, any small step in the right hemisphere self awakens, taking well the possibility of bringing to conscious life own contents of this hemisphere.

Recent advances in this research gives the brain the power to choose how to use the right and left hemispheres (Babiloni et al 2000), weighing in each case this convenience, this case possibly an activity for which commonly we accept as superior mind. In the brainstem found the ascending reticular system (ARS) which it is important in stimulating consciousness (Brain 1958).

If the Biofeedback is taken to a high level, the ARS sends signals to the cortex in order to reduce arousal (Brain 1995). ARS features provide the gateway to all forms of meditation, creative dreams and higher states of consciousness. The ARS is intimately connected to the limbic system and it seems that is the responsible for the phenomena associated with altered states of consciousness, states of euphoria, feeling of split consciousness, floating sensation, feeling white and golden light, and the like.

The hypothalamus is responsible for two highly integrated responses (Krout 2007):

- Sympathetic, fight or flight, reduces the skin resistance.

- Parasympathetic relaxation, increases the resistance of the skin.

Measures the resistance of the skin are (Solomon et al 1934):

Table 2 Resistance of the skin

State of consciousness	Resistance Ohm / cm2
Panic	50000
Normal	500,000-1,000,000
Dream	2,000,000

Using these measurements we can establish that a subject with stimulation reduced mental ARS will record a normal and vice versa, a subject very stimulated or anxious will record underweight ARS. In most individuals one hemisphere of the brain usually shows more active electricity than the other, but this difference tends to diminish and even disappear when the train starts in meditation.

There is broad agreement in the sense that the alpha brainwaves that represents a kind of synchronous understanding of cortex neurons. The premise of Biofeedback is that if a body can be associated with known mental state, you can control the body condition.

Physiological correlates of relaxation include (Benson et al 1978):

- Reduced consumption of O2

- Reduction of CO2 elimination

- Reduction of heart rate, respiratory rate, blood pressure, lactate blood, muscle tone, blood cortisol levels, etc.

- Increased perfusion of internal organs.

- Increasing the finger temperature.

- Increase skin resistance

- Increased intensity of alpha waves.

All this is called alertness and hypometabolic and is the fourth state of consciousness.

6.1 Uses of biofeedback

Uses of Biofeedback - for sake of simplicity uses will be under certain headings although some treatments are interchangeable. Instrumentation includes EEG, EMG, thermal.

A. Psychological - referring to treatment by a Psychologist or Psychiatrist.

ANXIETY REACTION - Overwhelming anxiety resulting from a situation which would not affect a person normally. For example, severe test anxiety where the person freezes up and blocks out all thoughts. Sexual performance anxiety where a male is unable to achieve an erection because of overwhelm- ing anxiety of performing. EMG, EEG, Thermal, SPR, relaxation exercises.

ESSENTIAL HYPERTENSION - High blood pressure with no apparent organic basis.

HEADACHES - Migraine, tension. EMG to lower tension and temp to open restricted vessels.

STRESS MANAGEMENT - Used in corporations with salespeople and executives. Using relaxation exercises to reduce stress monitor with EMG instrumentation.

POST VIETNAM STRESS SYNDROME - Used to combat the symptoms of insomnia, night mares, depression mood changes. Using EMG and thermal to walk thru the traumatic experience with the veteran allowing him to relive the experience and work through the feelings and reactions associated with the experience.

NEUROSIS - Similar to the approach used with anxiety reduction, sometimes anxiety becomes very severe, retards portion of personality. RELAXATION EXERCISES with EMG.

PHOBIA'S - Use of disensitization therapy with (SPR) or Thermal unit. Explain, use example of fear of snakes.

ALCOHOL AND DRUG ADDICTION - Used in conjunction with psychotherapy. Always a reason for the patient to need alcohol or drugs in order to cope. Use EEG to teach patient to obtain an altered state of awareness through mental power. You take away the alcohol and drugs and have to give the patient something in return. Also use EMG for tension and thermal for imagery enhancement. RELAXATION EXERCISES WHEN TENSE AND FEEL THE NEED FOR A DRINK OR DRUG.

DIABETES - A psychologist using a structural routine and relaxation techniques is able to help the diabetic become more aware of his system and in some cases cut down on the amount of insulin required. REQUIRES SCHEDULE OF EATING HABITS AND EXERCISE, AND/OR WORK.

CARDIOVASCULAR THERAPY - any biofeedback treatment of heart disease.

a. Essential Hypertension - covered in above section.

b. Cardiac Arrhythmias

1. Ectopic Thythm - premature contractions of the heart muscle.

2. Tochyarrhythmiss - irregular heart beat.

   11. NEUROMUSCULAR DISORDERS

a. Stroke

1. Hemiplegia - paralysis on one side of the body. Usually treatment is done with the J-53 dual site EMG, with an emphasis on muscle re-education. 4mm sensor placement.

2. Footdrop - EMG with small 4mm sensor placement.

b. Spinal cord injury

1. Paraplegia quadriplegia - again the use of EMG biofeedback looking for any unit of motor activity and working with that unit.

c. Torticollis - involves involuntary movement of neck and shoulder muscles. Using the J-53, Biofeedback involves toning down muscle tension on involuntary side and increase muscle tension on weak side.

d Peripheral Nerve Damage

1. Bell's palsy - EMG on facial paralysis area.

e Cerebral Palsy - EMG Biofeedback is used to work with posture and gait. The disease is a non-progressive neuromuscular disorder involving spasms and floridity. Involuntary head movements. EMG is used to monitor the patients ability to learn to control involuntary muscle movements.

12. Chronic Pain - EMG to relax pain area and temperature to warm the area.

13. Epilepsy - EEG to control brain waver pattern. When the patient becomes aware of his feelings before a seizure he immediately goes into his biofeedback exercises, altering his state of awareness and possibly curtailing the onset of the seizure. Through the use of biofeedback training many people suffering from epilepsy and taking medication (Phenobarbital) were able to cut down or stop the use of medication.

14. Parkinson's Desease - EMG biofeedback for the facial twitching.

15. Migraine Headaches - same as psychology section.

16. Speech Disorders - EMG Biofeedback

a. Stuttering - dysorthia (articulation disturbance, facial tension).

b. Tinnitus - (ringing or roaring in the ears) most loud in times of stress and tension. EMG and thermal to relieve tenseness; Subvocalization - talking to oneself while reading. EMG electrodes on the throat.

17. Gastrointestional Disorders

a. Urinary retention, EMG to control sphincter muscles and become aware of bloated bladder.

b. Functional Diarrhea - abdominal pain and bowel irregularity.

c. Encopresis - uncontrollable soiling, without bowel movements.

18. Dental Disorders

a. Bruxism - teeth clenching, usually done at night treatment usually entails a portable EMG instrument which wakes the patient up at night by a tone when he begins teeth grinding.

b. Myofacial Pain Dysfunction - (Pain and tenderness) in the jaw area, use EMG to relax muscles and cut down on the pain.

c. Orofacial Dyskinesia - involuntary facial movements-as a result of tardive dyskinesia. Again EMG to decrease muscle activity.

7 QUANTUM PHYSICS

What we are able to perceive with our five senses is not reality. Quantum physics has shown that space and time are illusions of perception. Our body cannot really be a reality if it does not occupy most of the space it seems to occupy; an experiment made at the University of Manchester revealed the shape of the interior of an atom is almost entirely empty space. The question then became how we could possibly make the world around us see us if this is the case (Russell *et al.,* 1993).

Our true consciousness does not exist in our brains or in our bodies, but this illusion of our individual bodies along with the misinformation of our true origins has manifested the idea that we all think independently from one another. With this understanding, it seems possible to scientifically explain telepathy, clairvoyance, spiritual mediums related to the transfer of information between sources without physical means of communication phenomena. But when we understand that there is a common spiritual bond between all things in the universe and that we are all part of a divine intelligence, this simple understanding fills all the holes in modern religions and predictions about the future and literally every occurrence of events (Russell *et al.,* 1993).

According to quantum physics, the physical world and its reality, it's just a recreation of the observed. We created the body and reality, as we create the experience of our world in its different manifestations dimensional. In its essential state (atomic or cosmic subquantum micro), the body is made of energy and information, not solid matter, this is only a meager level of perception. According to Tornell (2001), this is energy and information arising from the endless fields of energy and information spanning the entire universal creation.

When looking at the electronic microscope, we are looking at our microcosm, we then can see how the quantum particles manifest virtually as a symphony and intelligent orchestration at speeds much higher than the visible light, this quantum view also represents the immutability of our macrocosm. In this reality, each individual inhabits a reality that is beyond all change, as more deep within us without the knowledge of our three-dimensional or physical outer senses. There is a core of being, an energy field that creates immortality in nature, and manifests as the physical body. This core is the essential being or soul, primordial seed, which is contained in an atom called seed. We are seeds of eternal essence at this stage of quantum eternity.

This is the seed based on new paradigms posed by Planck, Maxwell, Faraday, Heisenberg, Schrödinger, Bohr, Einstein and Hawking, among many other pioneers of quantum mechanics. They understood that the way to see the world in their time was very wrong. Within the quantum paradigm, we are more than our physical body; our true self and personality are governed by the rules of the principle of cause and effect and are embedded into a body for the duration of human life. The field of human life is open and unlimited in its deeper quantum level, this means that we are immortal and timeless. Once we identify with the eternal reality that is consistent with quantum vision of the universe, we will enter the new paradigms of quantum consciousness. Everything that exists has a natural vibration to its atoms all the way up to the immensity of the universe to show a simple connection between land and our bodies.

It was Democritus, five centuries before the birth of Christ, who first suggested that the matter was not continuous, that is, which it was made by adhering small pieces infinity between yes, to which he called atoms ("indivisible" in Greek). Perhaps in those ancient times, science was not aware of the importance of this, it took centuries, not only to accept the theory, but already including his own name "atom". Much of quantum phenomenology is still theory, but some theories that have led to advances as the transistor, television, communications satellites, x-ray devices, computers, and ... the atomic bomb, with all the negative charge that entails, but also turn gave place to the cobalt bomb, scanners, nuclear power plants, etc. Current technical structure, it would not be possible or even imaginable without the knowledge and application of quantum mechanics.

Lord Rutherford of Nelson, a professor at Cambridge in the early years of the 20th century, was the first to describe the atom experimentally as a kind of planetary system dimensions "Unimaginably" tiny. In the center was what he called atomic nucleus, positively charged, and turned around a swarm of electrons, negatively charged, in a perfect balance. Subsequently he identified the positive charge of the nucleus was equal to the number of electrons that revolved around it, and in turn equal to the atomic number in the periodic table. Subsequent processing and measurements of this scheme, called Rutherford atom, evolved into new and more complex theory.

Max Planck in the Christmas gathering of German Physical Society, took the stand to outline what it would be a revolutionary proposal (James et al 2011). The light, which until then was supposed to be made of continuous waves without interruption, was now conveyed also in small packets of energy, with amounts energy package well defined: to this "minimal.

Plank called this energy package "quanto". This amount Power was proportional to the frequency of vibration light wave or electromagnetic emission that were, and thus it was unchanged for a given wavelength. Any amount of energy that is measured or observed, is always the sum quanta of composing, there then, as there is a continuous matter, continuous power; does not exist in the universe means quanto energy quanta or three and a half, the quanto is "the minimum that the universe dispatched "if we ask energy ".

Physical revolution that was implicit theory was not fully appreciated by Planck himself: among other things, just state what came to be called the "wave-particle duality", that is, the strange feature with light and electromagnetic waves sometimes behave as wave trains and sometimes intangible as particles with a mass and a specific load and something as the body and the spirit of things, to explain in language humanist. It's not a figure of speech, this dualistic behavior has been measured many times and depends precisely "how we want to see". To see it as particles exist a series of experiments and others see it as waves.

Based on Rutherford atom, and the quanta of Planck, the Danish physicist Niels Bohr, issued a new theory of the structure the atom, which in his honor is known as "atom Bohr '. This theory, first showed that electrons orbiting core on the model of Rutherford, would finish fall against the core itself because of the gravitational force exerted it. This could lead to a universal catastrophe, because nothing would exist otherwise (Bohr 1922).

Bohr suggested that electrons were carriers of mechanical energy due to its rotational speed, it was possible to measure quanta, as any energy should be an integral multiple quanto. Thus, each electron would have minimal power in the the core, and when gaining some energy, it would be possible to separate the core and perhaps even jump to another electron shell. In any case, the transition of an electron from a higher to lower energy states produces a quantum of light, and vice versa. Pauli (1940), with its call exclusion principle, then set the concept of how many electrons could be by layer, and how his mechanics. Here, we should note that the importance of gaining or losing electrons is significant in nature. For example, in a simplified form, the basic gas, the hydrogen has a proton (+) and a neutron (0) at its core, while around it has one electron (-). If hydrogen gains an electron and happens to have two it becomes another gas, helium, and in that conversion is issued an amount of energy that is the same as is similar to the hydrogen bomb, which is also similar to the sun and stars (nuclear fusion). In other words, the sun and stars are just large hydrogen bombs, gigantic factories converting hydrogen into helium. In contrast, if we would like to convert hydrogen helium, we should make one of the electrons that orbit the nucleus of helium jump its orbit and disappeared. This is not possible without providing some amount of energy, and the best way to achieve it is by bombarding we want ejected electron atomic particle to another to deviate it from his path (nuclear fission). Thus, it is possible convert other elements in accordance with the ancient tradition wanted alchemical, but the downside is indeed serious: the amount energy to be applied to convert any element to gold is much more expensive, even today, to acquire the same in any gold jewelry.

Thus, in the same way that the study of the atom led to the conclusion that the material (the surface of a table, a wall, and all you can imagine), is not continuous, but It is an intricate network of atoms linked together by the so-called force weak nuclear, power itself, in all its manifestations (Radiant, mechanical, etc.), is not continuous, but the sum of amounts "of quanta of energy.

The uncertainty principle Heisenberg (1958) had an enormous impact today since it is mitigated by modern concepts. Scientists, who once had the physics of Newton, led French Laplace to ensure that the universe was absolutely deterministic. This happened in the early nineteenth century. Laplace held that since the universe has rigorous laws and these know the state of the universe, these same laws would predict the future evolution of things. This thought was going a little further, by stating that these laws also exist for behavior human and therefore ruled all future possibilities of man. These ideas had a surprising success that reaches our days, but now the word determinism is replaced by that of "Destiny". "Everything is written", "we can not escape our destiny" "The future is inflexible," . This canceled at a stroke the first and most important gift that we received from God and that is none other than FREEDOM: freedom to live or die, freedom to love or hate, freedom to believe something or its opposite, etc. To which determinism answered with a technique of "a posteriori", because once the man action exercised his freedom by choosing something immediately answered: "Part of your destiny, as the universal laws governed at the time of your choice are really elected for you "Human freedom, from this perspective, was reduced a mere illusion.

Strong implications of the discoveries of Planck, Bohr, and many others not quoted in brevity, it was not fully appreciated until 1926, when the great German physicist Werner Heisenberg formulated his famous uncertainty principle. Heisenberg thought, correctly, that to predict the future of a particle was necessary to know the current speed and position and to study the present time particle that only one thing is essential. However, Heisenberg, concluded an unthinkable difficulty, if quanta content in light waves hits the particle, we will see its position, but we cannot deduce its speed. On the contrary, we have no way of detecting the passage of the particle by a point and another separated from the first, and measuring its speed does not allow us to know nothing about its position in space at any given time. Heisenberg demonstrated that one cannot know both the position and velocity of a particle in the future.

The implications of the work of Planck and Bohr had not been observed until Heisenberg enunciated his famous uncertainty principle, just as sure now that the implications of this principle has not yet been apprehended today all day, although there is much debate and are the subject of intense controversy. The Heisenberg uncertainty principle is obvious to note that governs for both particles and for the whole universe so is not possible to predict future events, since it is not measurable even the current state of play as necessary.

The philosophical importance derived from the uncertainty principle has indirectly reached various disciplines. E.g. in anthropology, he was certain that when a researcher visiting such a primitive tribe, he observed behavior of those Indians. That's not quite correct, because the presence of an Anthropologist modifies its behavior, and so much more as they are culturally different researcher and researched.

Heisenberg, with Schrödinger and Dirac formulated the called "mechanical Quantum ", which is to redefine, since it is not possible to know both the speed and position of a particle, the so-called "quantum state" which is a combination of both things at once. This does not leads to a single result is not predicted for each observation event, but a number of results are given in place calculating possible probabilities each (from the call wave function). A widely used example is the room where we are, for example, quantum mechanics can predict the millions of millions of different situations in the room space can occupy each of the molecules of air there, and likely to be met. One of these positions is that all the air in the room could focus on one of the upper corners, and we suffocate from lack of oxygen. This, however, is highly unlikely, but certainly happen in the time elapsed since now and 46 billion years in the future.

Einstein objected to the uncertainty principle since according to his famous phrase: "God does not play dice. " meant that he hated the idea of the final decision on which would likely among trillions of them possible at any given time, you were to chance. However, until today, experimentation supports fully the Heisenberg uncertainty principle.

Another interesting aspect of the theory of quantum mechanics is the confirmation of the wave-particle duality, relative concatenante with the Bohr theory of the atom comes to reinforce it.

One of the most famous and curious thought experiments the recent history of physics, is the Schrödinger's cat (Legget 1984). A cat is penned up in a steel chamber, along with the following device (which must be secured against direct interference by the cat): in a Geiger counter, there is a tiny bit of radioactive substance, so small, that perhaps in the course of the hour one of the atoms decays, but also, with equal probability, perhaps none; if it happens, the counter tube discharges and through a relay releases a hammer that shatters a small flask of hydrocyanic acid. If one has left this entire system to itself for an hour, one would say that the cat still lives if meanwhile no atom has decayed. The psi-function of the entire system would express this by having in it the living and dead cat (pardon the expression) mixed or smeared out in equal parts.

It is typical of these cases that an indeterminacy originally restricted to the atomic domain becomes transformed into macroscopic indeterminacy, which can then be resolved by direct observation. That prevents us from so naively accepting as valid a "blurred model" for representing reality. In itself, it would not embody anything unclear or contradictory. There is a difference between a shaky or out-of-focus photograph and a snapshot of clouds and fog banks. There is a fifty percent chance that each occurred but some must have consummated that the cat lives or is dead. Quantum physics is not so simple, as it follows from this simple proposition. The so-called Copenhagen Interpretation argues that there is a superposition of quantum states actually living / non-living cat, and it is absurd to wonder which of the two situations is the right until a determined observer. This is the called function collapse wave. At that time the cat was alive or dead, but only at that moment, to be an observer, "the universe would have decided ". Of course, the experiment is theoretical because it is not possible to completely isolate a showcase.

Now some of the issues raised in this mental exercise if present in the microscopic reality, include the one by the famous physicist David Bohm (Bohm & Hiley 1984), he says the situation is not possible described and that the cat is really dead or alive. To solve the problems of indeterminacy, he suggests a complex process of "hidden variables" that would eliminate conceptually. Finally not worth remembering the Interpretation Many Worlds of Everet III (1983), according to which the universe would unfold in two: one with a dead cat and a live, in which case we would be only one, but also in other without being aware of it-. This assumption is highly questionable since imply a doubling of the universe in every quantum process, thereby dramatically increases the complexity at each instant.

However, it is not entirely possible to banish for obvious reasons probative nature and even, in recent years, this has been the subject of depth studies as evidenced by the fact that it has been published in the journal "Scientific American" month of June 1994, a curious physical work known about. Until about 25 years ago, it was believed that the elementary particles of matter was the electrons, protons and neutrons, as constituent core atom. But experiments conducted on those produced collisions of protons with protons, electrons or protons, led to the conclusion that there were even more elementary particles in the matter. Indeed, the physics scientist Gell-Mann (1964) received in that year the Nobel Prize for the discovery of these particles, which he called "quarks". The study initiated by the quarks of Gell-Mann (1964) set the next surprise in quantum physics. Moreover, as good citizens of the quantum world, protons, neutrons, and quarks behave incomprehensible. A proton, for example, is composed of three quarks, two up quarks and one down quark (Generally quarks cannot be linked in varying amounts three). The resulting color of the sum of the three quark must be always the "white", ie the colors of them are mixed other to be canceled. Another feature is that the mass of the proton is less than the sum of the masses of the three quarks that comprise. Quarks, when bound into protons, neutrons, and other particles, they achieve such stability, for disintegrate and die "by itself" has not yet elapsed time precise since the universe exists for this to occur, and probably will not be in the future. However, when a quark is released in a collision between particles, it does not pass the half-life of a billionth of a second (Gell-Mann 1964). Another important concept in quantum theory is the concept of the spin. The spin is an essential feature of the particles and atomic sub-particles, and a brief description is to indicate the number of turns that should give those to observe all their properties. The spin is what defines and creates the differences between material particles and virtual (or forces) particles. For each particle there is its antiparticle, as Dirac predicted (Cooper, & Jennings 1986), this is particle with the same mass and opposite charge (including electric charge). For example, the antiparticle of the electron is the positively charged electron, or positron, which is produced naturally in certain types of radioactive decay.

There are four fundamental forces in the universe, of which all others derive. These are the electromagnetic force, the force weak nuclear, strong nuclear force and gravity. Its action is produced, it is believed theoretically, by the exchange of sub particles called respectively: virtual photons, bosons massive, gluons and gravitons. A grand unified theory is one that seeks to explain that these four forces are different manifestations of the same force that is the same force that ruled in the moment of the Big Bang (Georgi 1979).

7.1.1 Quantum consciousness

So only perceive with our five senses is not reality. Quantum physics has shown that space and time are illusions of perception, therefore, our body cannot really be a reality if they occupy the space.

An experiment to the University of Manchester revealed the shape of the interior of an atom. Scientists were surprised to discover the atom is almost entirely empty space. The question then became how I could possibly make the world around us see us.

Our true consciousness does not exist in our brains or in our bodies. But this illusion of our individual bodies along with the misinformation of our true origins has manifested the idea that we all think independently from one another. With this misunderstanding it seems possible to scientifically explain telepathy, clairvoyance, spiritual mediums related to the transfer of information between sources without physical means of communication phenomena. But when you understand that there is a common spiritual bond between all things in the universe and that we are all part of a divine intelligence there is unexplained phenomena. This simple understanding fills all the holes in modern religions chipped deja vu incarnation and predictions about the future and literally every occurrence of events or anomaly some experience.

According to quantum physics, the physical world and its reality, it's just a recreation of the observed. We created the body and reality, as we create the experience of our world in its different manifestations dimensional. In their essential state (atomic or cosmic subquantum micro), the body is made of energy and information, not solid matter, this is only a meager level of perception.

According to Tornell (2001), this energy and information are arising from the endless fields of energy and information spanning the entire universal creation. The mind and body, from the physical to the spiritual and multiple multidimensional manifestations are inseparably one unit that is "I am". This unit "I am", the studio will separate into two streams of experience.

The first experience as a subjective current, (thoughts, concepts, ideas, feelings, emotions, and desires). The current objective, the experience as a physical body, but yet on a deeper level, the two streams are in a single creative source (essence), and this is from where we really express and have our being. The biochemistry of the body is a product of quantum consciousness, feelings, emotions, thoughts and ideas, create reactions that sustain life in every cell. The perception of something, it seems like something automatic, but this is a learned phenomenon, if you change your perception, you change the experience of you, as this only has reality in your acting ability, be it on a conscious level, subconscious or supra conscious . And therefore this world. There impulses of intelligence that created your body in new ways every micro quantum moment. Which you are equivalent to the sum total of each quantum momentum to change these patterns change being.

Although each person appears as a separate and independent, in reality we are all connected to the patterns of universal intelligence, also called the absolute and merge with local terms like God. Our body is part of a universal body which in turn is part of the universal body. Our minds are part of the universal mind and in turn this is part of the universal mind.

Real-time, eternity exists as continuous present, is quantified eternity, timelessness is cut by us into pieces, or fragments of time we call days, hours, minutes, and seconds. What we call linear time is only a reflection of how we perceive the series of events or changes in our limited perceptual system is wrapped by the poor use of the brain system - cord - neuro.

The sequential time is given by the lack of ability to process all data experienced simultaneously, they would take what is called present continuous. Then the series of data sequences perceptions are processed in the brain according to its own processing power.

If you could perceive the changeless being, the time would be perceived and measured as we know, we have to learn to change the ability to process data and complexity of the process, to increase the level of consciousness (intelligence).

When looking at the electron microscope, we are looking at our microcosm, there we see how the quantum particles manifest virtually), a symphony and intelligent orchestration at speeds much higher than that of visible light, if we turn to heaven, we will see the immutability of all or macrocosm.

Each inhabits a reality that is beyond all change, as more deep within us without the knowledge of our three-dimensional or physical outer senses. There is a core of being, an energy field that creates immortality I like nature, and manifests as the physical body. This core is the being that is, the I am, the essential being or soul, primordial seed, which is contained in an atom called seed. We are seeds of eternity essential at this stage of quantum eternity.

This is the seed based on new paradigms posed by Max Planck, J. Clark Maxwell, Faraday, Heisenberg, Schrodinger, Bohr, Einstein, S. Hawking, among many other pioneers of quantum mechanics.

They understood that the way to see the world in its time, was very wrong, you are more apparent that your limited body, your self and your personality, (the current). The rules of the principle of cause and effect as we know, have gotten us into the volume of a body and the duration of human life. Actually the field of human life is open and unlimited in its deeper quantum level.

Being aware is to realize everything that happens around you. It is as if in the previous moment you take an action, before acting you ask yourself: Is it right what I do? and someone who "is not you" answers only "yes" or "no" (Hawnser, 1997). Our mind before ordering an action to the physical body, studies the elements it has to do with the action: knowledge, physical strength, skill, etc. Consider the risks, the consequences, the material benefits that give us the results of the act, also any loss and damage and / or suffering that the action may bring to our selves or third parties (Hawnser, 1997).

Edelman and Tonomi (2000) think that the whole universe is actually one living organism with full conscious awareness of self. The consciousness of our universe is responsible for the form and purpose that all matter assumes. Carl Jung (1981) found that there is a collective unconscious connected to all humans. This means that all humanity shares a single mind with one another. This is evident in the world through accounts of shared mythology and symbols. This collectivity is a global example of the unconscious mind of the human body in which billions of cells share a similar signal. Human consciousness is an electromagnetic energy field; this could explain many paranormal phenomena such as telepathy and clairvoyance that seem to probe this.

Most Western scientists assume that consciousness is produced in some form by the brain. There is of course some evidence for that position. There is evidence of common sense in our daily lives. When we drink too much alcohol or takes a hard hit to the head, we do not think clearly. We also have more sophisticated tests of the relationship between the brain and consciousness. In fact, all the theories of consciousness during the last century has been supported by psychologists who have been moving towards the materialism that characterized the nineteenth-century physics based on Newton's classical mechanics. These have been trying to show that consciousness is only the functioning of the physical brain. This materialistic psychology was supported by John Watson (1916), who wrote that psychology is a purely objective experimental branch of science that needs no consciousness in the same way that science does not need chemistry and physics. It is ironic that while Watson links psychology to the classical physical knowledge of Newtonian physics, there is overwhelming experimental evidence that the universe is related to quantum physics that could not be made without reference to consciousness.

John Lorber (1978) specialized in children with hydrocephalus, or water on the brain is a British neurologist. Children with this condition have an abnormal amount of cerebral spinal fluid accumulation in the cavities inside their brain compressing brain tissue that usually leads to mental retardation seizures, paralysis and blindness and if not treated to death. However, Lorber describes dozens of children and some adults with severe hydrocephalus but live normal lives. Indeed, in a sample of children with their cerebral space filled with ninety-five percent of spinal fluid in their skull leaving virtually no room for any brain tissue, half of them had a higher IQ than one hundred and thirty.

Some of the best evidence that consciousness can function independently of the brain come from near death experiences, profound experiences that some people report when they have been on the threshold of death. The near-death experiences are very short stories of people who have been clinically dead and then are resurrected or revived spontaneously after a brief interval with the memory of what they experienced during that period. According to Greyson (2010), many people with near death experiences reported vivid mental clarity exceptional sensory imagery and a clear memory of the experience and an experience that is more real, then in their daily lives.

Ageless body and mind time, we are immortal and timeless. Once we identify with the eternal reality and consistent with quantum omniverse vision, we will enter the new paradigms of quantum consciousness, this will expand their omniversales, radio, exponential and dimensional fractals.

Each particle omniverse, turns out to be a ghostly bunch of energy vibrating in an apparent void immense, (ether), the quantum field is not separate from us, "is us," that's where it all creates stars, galaxies, leptons, quarks, of all creation.

We are creating themselves to each nano-moment, in a huge capacity and creativity. The human body and all the whole cosmos is created and recreated every nano-moment, the body is a flowing body and potentiated by billions of years of intelligent experience. this intelligence is dedicated to monitor each nano-instant, constant change atropic and entropic, which takes place in each of ourselves, each cell is a terminal miniature connected to the cosmic computer or Omniversal mind we call all or God of all gods.

Morphogenesis is a scientific term to explain this very shaping of tissues organs and entire organisms (Gurwitsch 1915). Consciousness is the creative force of the entire universe. It has been given many names such as God Yahweh Krishna nature the field and divinity (Hick 1982).

The entire universe is in fact a single living conscious organism with complete awareness of the self.

The reason why it may seem difficult to comprehend this is because our understanding is typically limited by our language. When we hear the term conscious organism we tend to anthropomorphize its definition by giving it human qualities. We mistakenly look past what an organism truly is in the first place. The definition of an organism is any living thing capable of response to stimuli reproduction growth and development and maintenance of homeostasis as a stable whole our universe does. The consciousness of our universe is responsible for the form and   purpose   that   all   matter   assumes the conscious mind is to create a climate so one that has your personal identity and to what it does. The real thinking and then your subconscious mind. Well there's no way to test the subconscious mind it's equivalent of a tape player records behaviors and then to push a button placed behavior it's automatic. It's a very convenient thing because then we don't have to really. Learn all the time. Once you know what you can make a pattern. The problem is that the basic patterns of belief and behavior that are programmed subconscious mind came from our teachers primarily our parents our family and our community. Most people don't even understand how easily we are forced by our environment. Every person that we encounter every single situation that we're faced with every little word that said on television may seem too influential to our conscious minds but your unconscious is designed specifically to let every environmental signal influence you without your awareness. So the question is are we leaving conscious or unconscious lives. And now neuroscience is told us in the unfolding of our lives only five percent of our life is controlled by our conscious mind and ninety five percent of the time controlled by the subconscious. With programs from other people that were installed and the problem is these windows programs are playing we don't see them and the skeptics will sit there and say consciousness archetypes astrology. We create things with our hands not a mind. Archetypes aren't physical. They can influence me but when you think of the fact that we're only conscious of the small little fraction of our behavior what we don't realize is that entire countries entire civilizations that think they're free and independent but are unconsciously too afraid to be free and independent they will beg to be governed. And if they can do it themselves who do you think will consciously or unconsciously take that responsibility usually ends up being that strong masculine   and   in   this   archetypal   figure

Carl Young discovered that there is a collective unconscious connected to all humans (Jung 1936). Meaning that the whole of humanity shares a single mind with one another. This is evident in the world through accounts of shared mythology and symbols. This collectivity is a global example of the unconscious mind of the human body in which trillions of cells share a similar signal. This parasite called our false ego requires a continuous flow of sustenance to survive. Food fuel and any other form of sustenance is energy. Human consciousness is an electromagnetic field of energy. When this potential energy is utilized it then releases kinetic energy which is used to perpetuate the false ego.

This is why no matter how many civilizations rise and fall it is our collective consciousness that creates our governing apparatus not individual people. And after countless attempts you would imagine that people would realize that a physical retaliation may not be the solution. Yet here we are thousands of years later with technology that can clone D.N.A. vehicles that can break the sound barrier and probe the depths of space and science that can overcome almost any sickness. Yet we still fail to take notice to the importance of thoughts and consciousness. This is the very definition of insanity. And every single one of us is responsible for this psychic epidemic because we're killing the messenger and paying no attention to the message.

In this conscious living universe, there are no laws of nature just habits. There is nothing extra mile to the universe to enforce a law upon it. The illusion of a fixed law of nature is only the result of there being no need for that habit to be broken. When habits need to be broken to ensure the survival of the organism we see this event in nature and call it evolution. The collective mind shapes our evolution and a great example of this is the experiment done by John Carroll in one thousand nine hundred eighty eight. His team put lactose intolerant cells in an environment with only lactose for food. Under a law of nature every one of these lactose intolerant cells would have died. But surprisingly they all survived.

7.1.2 Earth frequency and consciousness

The earth's resonant frequencies are a result of its form. These frequencies are responsible for biological rhythms such as menstrual and circadian cycles as well as behavioral and emotional patterns. The frequencies are then picked up by the flora and fauna which are biological instruments that respond to the wave patterns (Adey,1993). The wave patterns resonate in the cranial structure of our head and converge in the center of our brain which is where we find the pineal gland (Adey,1993). The pineal gland is believed in many cultures to be the spiritual third eye responsible for intuition. Descartes called it the seat of the soul where mind and body meet each individual cell in our body receives an electromagnetic impulse from our central nervous system. They receive the very same impulse that was disseminated to every biological instrument from the earth. An explanation of our conscious universe has been attempted by religion science and philosophy. The neglect of biological nature from any organism causes illness a divorce from nature exile from the confounding of tongues. They're all symptoms not of a biblical god or deity   but   the   true   self.

The greatest and only threat to ourselves is a loss of self. The death of our divinity as we barrel through history with oceans of information yet barely a drop of wisdom. We have to understand how we lost herself the earth's resonant frequency starts at seven point eight three Hertz and ends with the seventh harmonic at forty three point two Hertz correlating with the seven shockers, ultimately the greatest discovery of our earth as its consciousness a visible attribute of consciousness is an energetic field that governs the shaping of organisms (Penrose & Hameroff 2011).

8 REFERENCES

Adrian E and Matthews B (1934). The Berger Rhythm; potential Changes from the occipital lobes in man. Brain; 57(1): 355-384.

Aveline, M. (1992). The use of audio and videotape recordings of therapy sessions in the supervision and practice of dynamic psychotherapy. British Journal of Psychotherapy, 8(4), 347-358.

Adeney, F. S. (1988). Transpersonal psychology: Psychology and salvation meet. In K. Hoyt and the Spiritual Counterfeits Project (Eds.), *The new age rage* (pp. 107-127). Old Tappan, NJ: Fleming H. Revell.

Ader, R., Felton, D. L., & Cohen, N. (Eds.). (2000). *Psychoneuroimmunology* (3rd ed.) (Vols. 1-2). New York: Academic.

Adey, W. R. (1993). Biological effects of electromagnetic fields. Journal of cellular biochemistry, 51, 410-410.

Alcañiz, M., Botella, C., Banos, R. M., Zaragoza, I., & Guixeres, J. (2009). The Intelligent e-Therapy system: a new paradigm for telepsychology and cybertherapy. British Journal of Guidance & Counselling, 37(3), 287-296.

American Transpersonal Association. (1969). Statement of purpose. *Journal of Transpersonal Psychology, 1*, i.

Anthony, D., Ecker, B., & Wilber, K. (Eds.). (1987). Spiritual choices: The problem of recognizing authentic paths to inner transformation. New York: Paragon House.

Astin, J. A., & Shapiro, D. H. (1997). Measuring the psychological construct of control: Applications to transpersonal psychology. *Journal of Transpersonal Psychology, 29*, 63-72.

Babiloni, F., Cincotti, F., Lazzarini, L., Millan, J., Mourino, J., Varsta, M., ... & Marciani, M. G. (2000). Linear classification of low-resolution EEG patterns produced by imagined hand movements. Rehabilitation Engineering, IEEE Transactions on, 8(2), 186-188.

Barad, K. (2007). Meeting the universe halfway: Quantum physics and the entanglement of matter and meaning. Duke University Press.

Barr, F.E. (1984) What Is Melanin? Unpublished article available by request from the Institute for the Study of Consciousness, 2924 Benvenue Ave., Berkeley CA 94705

Becker, RO, & Selden, G. (1985). The Body Electric: Electromagnetism and the foundation of life (pp 364-p.). New York: Quill.

Beloff, J. (1993). *Parapsychology: A concise history.* New York: St. Martin's Press.

Bem, D. J., Palmer, J., & Broughton, R. S. (2001). Updating the ganzfeld database: A victim of its own success? *Journal of Parapsychology, 65,* 207-218.

Bennett JM, Catovsky D, Daniel MT, Flandrin G, Galton DAG and Gralnik HR. Lithotripsy (1985). Health and Public Policy Committee, American College of Physicians. Annals of Internal Medicine 103(4): 626-629.

Benson, H., Dryer, T., & Hartley, L. H. (1978). Decreased [Vdot] O2 Consumption during Exercise with Elicitation of the Relaxation Response. Journal of human stress, 4(2), 38-42.

Benjamin Jr, L. T. (2007). A brief history of modern psychology. Blackwell Publishing.

Boals, G. F. (1978). Toward a cognitive reconceptualization of meditation. *Journal of Transpersonal Psychology, 10*, 143-182.

Bohm, D. (1990). A new theory of the relationship of mind and matter. Philosophical psychology , three (2-3), 271-286.

Bohm, D. (2002). Wholeness and the implicate order (Vol. 10). Psychology Press.

Botella C, Garcia-Palacios A, Banos RM and Quero S (2009). Cybertherapy: Advantages, Limitations and Ethical issues, PsychNology Journal; 7(1): 77-100

Boucouvalas, M. (1980). Transpersonal psychology: A working outline of the field. *Journal of Transpersonal Psychology, 12*, 37-46.

Berger, H. (1929). Über das elektrenkephalogramm des menschen. European Archives of Psychiatry and Clinical Neuroscience, 87(1), 527-570.

Bloch, M., & Parry, J. (1982). Death and the Regeneration of Life. Cambridge University Press.

Bohr, N. (1922). On the quantum theory of line-spectra. Bianco Lunos Bogtrykkeri.

Bohm, D., & Hiley, B. J. (1984). Measurement understood through the quantum potential approach. Foundations of Physics, 14(3), 255-274.

Bonny, HL (1975). Music and consciousness. Journal of music therapy, 12(3): 121-135.

Braud, L. W., & Braud, W. G. (1974). Further studies of relaxation as a psi-conducive state. *Journal of the American Society for Psychical Research, 68*, 229-245.

Braud, W. G. (2003). Distant mental influences: Its contributions to science, healing, and human interactions. Charlottesville, VA: Hampton Roads Publishing.

Braud, W. G. (1997). Parapsychology and spirituality: Implications and intimations. In C. T. Tart (Ed.), *Body, mind, spirit: Exploring the parapsychology of spirituality* (pp. 135-152). Charlottesville, VA: Hampton Roads.

Braud, W. G. (2001). Experiencing tears of wonder-joy: Seeing with the heart's eye. *Journal of Transpersonal Psychology, 33*, 99-112.

Brain, R. (1958). The physiological basis of consciousness. Brain, 81, 426-455.

Brown, D., & Engler, J. (1980). The stages of meditation: A validation study. *Journal of Transpersonal Psychology, 12*, 143-192.

Brown, D., & Pedder, J. (1979). Introduction to psychotherapy: An outline of psychodynamic principles and practice (Vol. 190). Tavistock Publications.

Brown, MF (1994). Share Spirits Who Owns What? Reflections on Commodification and Intellectual Property in New Age America. POLAR: Political and Legal Anthropology Review , 17 (2), 7-18.

Bucke, M. A. (1901/1969). *Cosmic consciousness: A study in the evolution of the human mind.* New York: E. P. Dutton. (Original work published 1901)

Bucke, RM (2009) Cosmic consciousness: A study in the evolution of the human mind . Courier Corporation.

Butts, R. (2003a). The personal sessions: Book 1 of the deleted Seth material. Manhasset, NY: New Awareness Network.

Butts, R. (2003b). The personal sessions: Book 2 of the deleted Seth material. Manhasset, NY: New Awareness Network.

Budzynski, T., Jordy, J., Budzynski, H. K., Tang, H. Y., & Claypoole, K. (1999). Academic performance enhancement with photic stimulation and EDR feedback. Journal of Neurotherapy, 3(3-4), 11-21.

Caddy, E. (2007). Opening Doors Within: 365 Daily Meditations from Findhorn. Findhorn Press.

Cade, C. M. (1989). The awakened mind: biofeedback and the development of higher states of awareness. Element Books Limited.

Cade, C. Maxwell, and Nona Coxhead. 1979. The Awakened Mind: Biofeedback and the Development of Higher States of Awareness. New York: Delacorte Press.

Cahn, B. R., & Polich, J. (2006). Meditation states and traits: EEG, ERP, and neuroimaging studies. Psychological bulletin, 132(2), 180.

Carlat, D. J. (1989). Psychological motivation and the choice of spiritual symbols: A case study. *Journal of Transpersonal Psychology, 21*, 139-148.

Capra, F. (1982). The turning point: Science, society, and the rising culture. New York: Bantam.

Charman, R. A. (2000). Placing healers, healees, and healing into a wider research context. The Journal of Alternative and Complementary Medicine, 6(2), 177-180.

Clare, A. W., & Thompson, S. (1981). Let's talk about me: a critical examination of the new psychotherapies. British Broadcasting Corp..

Cooper, E. D., & Jennings, B. K. (1986). On the role of antiparticles in Dirac phenomenology. Nuclear Physics A, 458(4), 717-724.

Chandley, M (1986), A psychological investigation of the development of the process in personality function mediumistic, PhD dissertation, International College

Crick, F., & Koch, C. (2003). A framework for consciousness. Nature neuroscience, 6(2), 119-126.

Cunningham, PF (2012). The content-source research problem in modern mediumship. Journal of Parapsychology, 76 (2), 295.

d'Aquili E (1983). The Myth-Ritual Complex: A Biogenetic Structural Analysis, Zygon 18(1): 247-269.

d'Aquili, E (1986). Myth, Ritual and the Archetypal Hypothesis, Zygon 21(2): 141-160.

Damasio, A. R. (1989). Time-locked multiregional retroactivation: A systems-level proposal for the neural substrates of recall and recognition. Cognition, 33(1), 25-62.

Dehaene, S., & Naccache, L. (2001). Towards a cognitive neuroscience of consciousness: basic evidence and a workspace framework. Cognition, 79(1), 1-37.

De Chardin, PT (1965). The phenomenon of man (Vol. 383). New York, NY, USA :: Harper & Row.

Dennett, D. C. (1993). Consciousness explained. Penguin UK.

Dharmananda, S. (2003). Gallnuts and the uses of Tannins in Chinese Medicine. ITM.

Dumit, J. (1995). Brain-mind machines and American technological dream marketing: towards an ethnography of cyborg envy. The cyborg handbook, 347-362.

Edelman, G. M. (1993). Neural Darwinism: selection and reentrant signaling in higher brain function. Neuron, 10(2), 115-125.

Edelman, GM, & Tononi, G. (2000). A Universe of Consciousness: How Matter Becomes Imagination. Basic books.

Efran, J. S., & Clarfield, L. E. (1992). Constructionist therapy: Sense and nonsense. Therapy as social construction, 200-217.

Emotiv, 3D-Brain (2016), https://emotiv.com/store/product_74.html Accessed date: January 10, 2016.

Everett III, H. (1963). Generalized Lagrange multiplier method for solving problems of optimum allocation of resources. Operations research, 11(3), 399-417.

Euler, C. (1986). Brain stem mechanisms for generation and control of breathing pattern. Comprehensive Physiology.

Faber, P. L., et al. "Scalp and intracerebral (LORETA) theta and gamma EEG coherence in meditation." International Society for Neuronal Regulation, Winterthur, Switzerland (2004).

Fetter, M. S. (2009). The Internet: Backbone of the World Wide Web. Issues in mental health nursing, 30(4), 281-282.

Ferrer, JN (2002). Revisioning transpersonal theory: A participatory vision of human spirituality. Suny Press.

Fischbach, M. (2002). Rare genetic diseases—new opportunities and challenges through biotechnological progress and scientific knowledge. European Journal of Paediatric Neurology, 6, A71-A75.

Fodor, J. A. (1992). A theory of the child's theory of mind. Cognition, 44(3), 283-296.

Flanagan, J. C. (1954). The critical incident technique. Psychological bulletin, 51(4), 327.

Flexihub (2016), Flexihub software, http://www.flexihub.com/ Accessed date: January 15, 2016.

Freud, S., & Strachey, JE (1964). The standard edition of the complete psychological works of Sigmund Freud.

Freud, S. (1978). The interpretation of dreams . Hayes Barton Press.

Hawnser, PE (1997). The Answer. Mexico: Editorial Diana.

Gell-Mann, M. (1964). A schematic model of baryons and mesons. Physics Letters, 8(3), 214-215.

Georgi, H. (1979). Towards a grand unified theory of flavor. Nuclear Physics B, 156(1), 126-134.

Gilula, M. F., & Kirsch, D. L. (2005). Cranial electrotherapy stimulation review: a safer alternative to psychopharmaceuticals in the treatment of depression. Journal of Neurotherapy, 9(2), 7-26.

Goleman D and Davidson RJ (1979). Consciousness: Brain, states of consciousness, and mysticism. Harper, New York,

Gliem, F., Dehmel, G., Musmann, G., Tuerke, C., Krupstedt, U., & Kugel, R. P. (1976). The onboard computers of the Helios magnetometer experiments E 2 and E 4. Raumfahrtforschung, 20, 16-19.

Green E and Alyce G. Beyond biofeedback (1977). Delacorte, New York.

Green E, Alyce G and Walters ED (1999). Voluntary control of internal states: Psychological and physiological. Subtle energies and energy medicine 10(1): 71-88.

Green, E. E., & Green, A. M. (1971). On the meaning of transpersonal: Some metaphysical perspectives. *Journal of Transpersonal Psychology, 3*, 27-46.

Greenland, S., Sheppard, AR., Kaune, WT, Poole C, Kelsh MA and Childhood (2000) Leukemia-EMF Study Group. A pooled analysis of magnetic fields, wire codes, and childhood leukemia. Epidemiology, 11(6): 624-634.

Greyson, B. (2010). Implications of near-death experiences for a postmaterialist psychology. Psychology of Religion and Spirituality , 2 (1), 37.

Grinspoon L and James BB. Psychedelic drugs reconsidered (1979). Basic Books, New York,

Grof, S. (1973). Theoretical and empirical basis of transpersonal psychology and psychotherapy: Observations from LSD research. Journal of Transpersonal Psychology, 5 (1), 15-53.

Grof, S. (1994). Transpersonal psychology: birth, death and transcendence in psychotherapy. Editorial Kairos.

Grof, Stanislav (1988). The Adventure of Self-Discovery. New York: State University of New York Press.

Grof, S. (2000). Psychology of the Future: Lessons from modern consciousness research. Albany, NY: State University of New York Press.

Groth-Marnat, G. (2009). Handbook of psychological assessment (5th ed.). Hoboken, NJ: John Wiley & Sons.

Gurwitsch, A. (1915). On practical vitalism. American Naturalist, 763-770.

Hall CS and Vernon JN (1973). A Primer of Jungian Psychology. Mentor Books, New York.

Hafner, R. J. (1982). Psychological treatment of essential hypertension: a controlled comparison of meditation and meditation plus biofeedback. Biofeedback and Self-regulation, 7(3), 305-316.

Hardt, JV, & Kamiya, J. (1978). Anxiety change through alpha electroencephalographic feedback seen only in high anxiety subjects. Science, 201 (4350), 79-81.

Hämäläinen, M., Hari, R., Ilmoniemi, R. J., Knuutila, J., & Lounasmaa, O. V. (1993). Magnetoencephalography—theory, instrumentation, and applications to noninvasive studies of the working human brain. Reviews of modern Physics, 65(2), 413.

Hameroff, S. R. (1994). Quantum coherence in microtubules: A neural basis for emergent consciousness?. Journal of Consciousness Studies, 1(1), 91-118.

Hatch, G. P. and Stelter R. E. Magnetic design considerations for devices and particles used for biological high-gradient magnetic separation (HGMS) systems (2001). Journal of Magnetism and Magnetic Materials, 225(1): 262-276.

Hawnser PE (1977). The Answer. Editorial Diana, Mexico.

Health and Public Policy Committee, American College of Physicians. (1985). Annals of Internal Medicine, 102, 709-715.

Heisenberg, W. (1958). Physics and philosophy: The revolution in modern science.

Hersen, M., & Turner, S. M. (Eds.). (2003). Diagnostic interviewing (3rd ed.). New York, NY: Kluwer Academic/Plenum Publishers.

Hersen, M., & Van Hasselt, V. B. (1998). Basic interviewing: A practical guide for counselors and clinicians. Mahwah, NJ: Erlbaum.

Hess, D. J. (1996). Technology and alternative cancer therapies: an analysis of heterodoxy and constructivism. Medical anthropology quarterly, 10(4), 657-674.

Honorton C (1970). Tracing ESP through altered states of consciousness, Psychic Magazine 2(18):78

Howard, C. E. (1986). A comparison of methods for reducing stress among dental students. Journal of dental education, 50(9), 542-44.

Howard, W. R. (1992). Patient Applied Tooth Whiteners are They Safe, Effective with Supervision?. The Journal of the American Dental Association, 123(2), 57-60.

Hick, J. (1982). God has many names. Westminster John Knox Press.

Huang, T. L., & Charyton, C. (2008). A comprehensive review of the psychological effects of brainwave entrainment. Altern Ther Health Med, 14(5), 38-50.

Hudson, TJ (1904) The Law of Psychic Phenomena: A Working Hypothesis for the Systematic Study of Hypnotism, Spiritism, Mental Therapeutics, Etc AC McClurg & Company

Hutchison, M. (1986). Megabrain: New tools and techniques for brain growth and mind expansion. William Morrow & Company.

Hyman, R. (1985). The ganzfeld psi experiment: A critical appraisal. Journal of Parapsychology.

Hu, H., & Wu, M. (2012). Scientific GOD: Michael Persinger & the GOD Experiments. Scientific GOD Journal, 3(10).

Hutchison M (1986). Megabrain: New tools and techniques for brain growth and mind expansion. Beech Tree Books.

Huxley, A., & Bradshaw, D. (1945). The perennial philosophy (p. 55). New York: Harper

Irwin, HJ, & Watt, CA (2007). An introduction to parapsychology . McFarland.

Ivey, A. E., Ivey, M. B., & Zalaquett, C. P. (2009). Intentional interviewing and counseling: Facilitating client development in a multicultural society. Belmont, CA: Wadsworth.

James, W. (1985). The Varieties of Religious Experience (Vol. 15). Harvard University Press.

James, W. (1985). The varieties of religious experience (Vol. 13). Harvard University Press.

James, J., Steinhauser, T., Hoffmann, D., & Friedrich, B. (2011). One Hundred Years at the Intersection of Chemistry and Physics: The Fritz Haber Institute of the Max Planck Society 1911-2011. Walter de Gruyter.

Jonas WB and Levin JS (1999). Essentials of complementary and alternative medicine. Lippincott Williams & Wilkins, Philadelphia, Pa, USA.

Josephson, B. D. (1962). Possible new effects in superconductive tunnelling. Physics letters, 1(7), 251-253.

Joyce, M., & Siever, D. (2000). Audio-visual entrainment program as a treatment for behavior disorders in a school setting. Journal of Neurotherapy, 4(2), 9-25.

Jung, C. G. (1936). The concept of the collective unconscious. Collected works, 9(1), 42.

Jung, CG (1981). The archetypes and the collective unconscious (No. 20). Princeton University Press.

Kasamatsu, A., & Hirai, T. (1966). An electroencephalographic study on the Zen meditation (Zazen). Folia Psychiatrica et Neurologica Japonica, 20, 315-336

Kamiya J. (1969) Operant monitoring of the EEG alpha rhythm and some of Its Reported effects on consciousness, In C. T. Tart (Ed.), Altered states of consciousness (pp. 519-529). Anchor Books, Garden City, N. Y.

Kiesler, D. J., & Van Denburg, T. F. (1993). Therapeutic impact disclosure: A last taboo in psychoanalytic theory and practice. Clinical Psychology & Psychotherapy, 1(1), 3-13.

King, C. (2012). Entheogens, the Conscious Brain and Existential Reality: Part 1. Journal of Consciousness Exploration & Research, 3(6).

Kirlian, S. D., & Kirlian, V. (1973). Photography by means of high-frequency currents. Galaxies of life. New York: Interface.

Kilner, W. J. (1965). The human aura. Citadel Press.

Klimo, J. (1998). Channeling: Investigations on receiving information from paranormal sources . North Atlantic Books.

Koch, C., Poggio, T., & Torre, V. (1983). Nonlinear interactions in a dendritic tree: localization, timing, and role in information processing. Proceedings of the National Academy of Sciences, 80(9), 2799-2802.

Koch, C., & Crick, F. (1994). CT Some Further Ideas Regarding the Neuronal Basis of Awareness. Large-scale neuronal theories of the brain, 93.

Kottler, MJ (1974). Alfred Russel Wallace, the origin of man, and spiritualism. Isis, 145-192.

Kroger, W. S., & Schneider, S. A. (1959). An electronic aid for hypnotic induction: A preliminary report. International Journal of Clinical and Experimental Hypnosis, 7(2), 93-98.

Krout, R. E. (2007). Music listening to facilitate relaxation and promote wellness: Integrated aspects of our neurophysiological responses to music. The arts in Psychotherapy, 34(2), 134-141.

Krippner, S. (2000). The epistemology and technologies of shamanic states of consciousness. Journal of Consciousness Studies, 7(11-12), 93-118.

Laudon, K. C., & Traver, C. G. (2011). E-Commerce 2011: Business, technology.

Lisman, J. E., & Idiart, M. A. (1995). Storage of 7+/-2 short-term memories in oscillatory subcycles. Science, 267(5203), 1512-1515.

LaBerge, S., & Rheingold, H. (1990). Exploring the world of lucid dreaming (p. 24). New York: Ballantine Books.

Lawrence J. Alpha Brain Waves. Nash Publishing. Los Angeles, 1072.

Le Scouranec, R. P., Poirier, R. M., Owens, J. E., & Gauthier, J. (2001). Use of binaural beat tapes for treatment of anxiety: a pilot study of tape preference and outcomes. Alternative therapies in health and medicine, 7(1), 58.

Lesh TV (1970). Meditators Zen and the Development of Empathy in Counselors, Journal of Humanistic Psychology, 10(1):39-74.

Leggett, A. J. (1984). Schrödinger's cat and her laboratory cousins. Contemporary Physics, 25(6), 583-598.

Libet, B., Freeman, A., & Sutherland, K. (2000). The volitional brain: Towards a neuroscience of free will (Vol. 6). Imprint Academic.

Liboff AR (1985). Geomagnetic cyclotron resonance in living cells. Journal of Biological Physics , 13 (4), 99-102.

Lorber, J. (1978, January). Is Your Brain Really Necessary. In Archives of Disease in Childhood (Vol. 53, No. 10, pp. 834-834). MED ASSOC BRITISH HOUSE, Tavistock Square, London, England WC1H 9JR: BRITISH JOURNAL MED GROUP PUBL.

Ludwig A. Altered states of consciousness, Archives of general Psychiatry 1966: 15(3) 225-234.

Magnus O and Van der Holst M. Zeta (1987) waves: a special type of slow delta waves. Electroencephalography and clinical neurophysiology 67 (2):140-146.

Maslow, AH (1969). Various meanings of transcendence. Journal of Transpersonal Psychology, 1 (1), 56-66.

Magnus, O., & Van der Holst, M. (1987). Zeta waves: a special type of slow delta waves. Electroencephalography and clinical neurophysiology, 67 (2), 140-146.

Maslow, AH (1969). Various meanings of transcendence. Journal of Transpersonal Psychology, 1 (1), 56-66.

Masters R and Jean H (1966). The Varieties of Psychedelic Experience. Delta Books, New York.

Marshall, I. N., & Zohar, D. (1997). Who's Afraid of Schrödinger's Cat?: All the New Science Ideas You Need to Keep Up with the New Thinking (p. 402).

Miller NE. Biofeedback and visceral learning (1978). Annual Review of Psychology; 29 (1), 373-404.

Mitchell, E. D. & White, J. (Eds.). (1974). *Psychic exploration: A challenge for science.* New York: Putnam.

McGinn C (1991) The Problem of Consciousness: Essays Toward a Resolution, Blackwell US.

Moore, J. P., Trudeau, D. L., Thuras, P. D., Rubin, Y., Stockley, H., & Dimond, T. (2000). Comparison of alpha-theta, alpha and EMG neurofeedback in the production of alpha-theta crossover and the occurrence of visualizations. Journal of Neurotherapy, 4(1), 29-42.

Monroe, R. A. (1977). Journeys out of the body (Vol. 79). Main Street Books.

Monroe, R. (1982). The Hemi-Sync process. Monroe Institute Bulletin,# PR31380H. Nellysford, VA.

Moorcroft, W. H., & Belcher, P. (2003). Understanding sleep and dreaming (pp. 168-169). York: Kluwer Academic/Plenum Publishers.

Myers, FW (1895). The subliminal self. In Proceedings of the Society for Psychical Research (Vol. 11, pp. 334-593).

Musso, JR (1994). THE IMPORTANCE OF parapsychology for psychology and psychoanalysis. Journal of Psychology Paranormal Argentina , 5 (3).

Nakagawa, K. (1976). Magnetic field deficiency syndrome and magnetic treatment. Japan Med J, 2745, 24-32.

Noton, D. (2000). Migraine and photic stimulation: report on a survey of migraineurs using flickering light therapy. Complementary Therapies in Nursing and Midwifery, 6(3), 138-142.

NOMURA, T., HIGUCHI, K., YU, H., SASAKI, S. I., KIMURA, S., ITOH, H., ... & KAWAI, K. (2006). Slow-wave photic stimulation relieves patient discomfort during esophagogastroduodenoscopy. Journal of gastroenterology and hepatology, 21(1), 54-58.

Nunez, P. L., & Srinivasan, R. (2006). Electric fields of the brain: the neurophysics of EEG. Oxford university press.

Ornstein, RE (1973) The nature of human consciousness: A book of readings . WH Freeman.

Ornstein RE (1972). The psychology of consciousness, WH Freeman, New York.

Olmstead, R. (2005). Use of auditory and visual stimulation to improve cognitive abilities in learning-disabled children. Journal of Neurotherapy, 9(2), 49-61.

Ornstein RE (1972). The psychology of consciousness, WH Freeman, New York.

Ossebaard, H. C. (2000). Stress reduction by technology? An experimental study into the effects of brainmachines on burnout and state anxiety. Applied psychophysiology and biofeedback, 25(2), 93-101.

Owen, IM, & Sparrow, M. (1976). Conjuring up Philip: An adventure in psychokinesis. HarperCollins Publishers.

Padmanabhan, R., Hildreth, A. J., & Laws, D. (2005). A prospective, randomised, controlled study examining binaural beat audio and pre-operative anxiety in patients undergoing general anaesthesia for day case surgery*. Anaesthesia, 60(9), 874-877.

Pahnke DR. Good Friday experiment: A long-term follow-up and methodological critique. Journal of Transpersonal Psychology 1991; 23 (1):1-28.

Penfield W and Perot P. The brain record of auditory and visual experience a final summary and discussion, Brain 1963: 86 (4); 595-696.

Pauli, W. (1940). The connection between spin and statistics. Physical Review, 58(8), 716.

Penrose, R. (1994). Shadows of the Mind (Vol. 4). Oxford: Oxford University Press.

Penrose, R., & Hameroff, S. (2011). Consciousness in the universe: Neuroscience, quantum space-time geometry and Orch OR theory. Journal of Cosmology, 14, 1-17.

Persinger, M. A. (1983). Religious and mystical experiences as artifacts of temporal lobe function: a general hypothesis. Perceptual and motor skills, 57(3f), 1255-1262.

Persinger, M. A., Tiller, S. G., & Koren, S. A. (2000). Experimental simulation of experience and a elicitation haunt of activity by paroxysmal electroencephalographic transcerebral complex magnetic fields: induction of a synthetic ghost. Perceptual and Motor Skills, 90(2), 659-674.

Persinger, M. A., Saroka, K. S., Koren, S. A., & St-Pierre, L. S. (2010). The electromagnetic induction of mystical and altered states within the laboratory. J. Conscious. Explore Res, 1(7), 808-830.

Penrose, R. (1994). Shadows of the Mind (Vol. 4). Oxford: Oxford University Press.

Poston, J. M. & Hanson, W. E. (2010). Meta-analysis of psychological assessment as a therapeutic intervention. Psychological Assessment, 22(2), 203–212.

Pribram, KH (1991). Brain and perception: holonomy and structure in figural processing . Psychology Press

Rank, O. (1929). The trauma of birth. Courier Corporation.

Reich, W. (1951). Cosmic Superimposition: Man's Orgonotic Roots in Nature. Wilhelm Reich Foundation.

Roberts, J. (2012). The Individual and the Nature of Mass Events (A Seth Book). Amber-Allen Publishing.

Rodgers, R. (2001). Handbook of diagnostic and structured Interviewing. New York, NY: Guilford Press.

Rogo, DS (1975). Parapsychology: A century of inquiry. Taplinger Publishing Company.

Russell RJ, Murphy N. and Isham C J (1993). Quantum cosmology and the laws of nature: scientific perspectives on divine action. Vatican Observatory, Italy,

Sabourin, M. E., Cutcomb, S. D., Crawford, H. J., & Pribram, K. (1990). EEG correlates of hypnotic susceptibility and hypnotic trance: spectral analysis and coherence. International Journal of Psychophysiology, 10(2), 125-142.

Shapiro Jr, D. H. (2008). Meditation: Self-regulation strategy and altered state of consciousness. Transaction Publishers.

Shapiro, S. I., Lee, G. W., & Gross, P. L. (2002). The essence of transpersonal psychology: Contemporary views. *The International Journal of Transpersonal Studies, 21,* 19-32.

Shakti   (2016),   Shakti-Neuromagnetic   signal   generator, https://www.shaktitechnology.com/winshakti/rotating/ Accessed date: January 8, 2016.

Sculthorpe L and Persinger MA (2003). Does phase-modulation of applied 40-Hz transcerebral magnetic fields affect subjective experiences and hypnotic induction?. Perceptual and motor skills, 97(3): 1031-1037.

Sejnowski, T. J. (1986). Open questions about computation in cerebral cortex. Parallel distributed processing, 2, 372-389.

Singer, JL, & Streiner, BF (1966). Imaginative content in the dreams and fantasy play of blind and sighted children. Perceptual and Motor Skills , 22 (2), 475-482.

Solomon, A. P., & ENTRESS, T. L. (1934). GALVANIC SKIN REFLEX AND BLOOD PRESSURE REACTIONS IN THE PSYCHONEUROSES*. The Journal of Nervous and Mental Disease, 80(2), 163-182.

Solomon, G. D. (1985). Slow Wave Photic Stimulation in the Treatment of Headache-a Preliminary Report. Headache: The Journal of Head and Face Pain, 25(8), 444-446.

Sommers-Flanagan, J., & Sommers-Flanagan, R. (2008). Clinical interviewing. Hoboken, NJ: John Wiley & Sons.

Smith H (1964). Have religious import drugs do? The Journal of Philosophy; 61(1): 517-530.

Smith, H. (1965). The religions of man . New York: Harper & Row

Siever, D. (2000). The rediscovery of audio-visual entrainment technology. Unpublished manuscript.

Sonty, N. (2003). Biofeedback as an Adjunct in Rehabilitation Medicine. Alternative Medicine and Rehabilitation: A Guide for Practitioners, 197.

Smith, H. (1965). The religions of man . New York: Harper & Row

Suzuki, L. A., Ponterotto, J. G., & Meller, P. J. (2008). Handbook of multicultural assessment: Clinical, psychological, and educational applications. San Francisco, CA: Jossey-Bass.

Sutich, AJ (1969). Some Considerations Regarding transpersonal psychology. Journal of Transpersonal Psychology, 1 , 11-20.

Szasz, T. (1988). Schizophrenia: The sacred symbol of psychiatry. Syracuse University Press.

Takacs, B. (2005). Special education and rehabilitation: teaching and healing with interactive graphics. Computer Graphics and Applications, IEEE, 25(5), 40-48.

Takahashi, T., Murata, T., Hamada, T., Omori, M., Kosaka, H., Kikuchi, M., et al.. (2005). Changes in EEG and autonomic nervous activity during meditation and their association with personality traits. International Journal of Psychophysiology, 55,199-207.

Tart, CT (1975). States of consciousness (p. 206). New York: EP Dutton.

Tart CT (1971). Scientific foundations for the study of altered states of consciousness. Journal of Transpersonal Psychology 3(1): 93-124.

Tart CT (1976). The basic nature of altered states of consciousness: A systems approach. Journal of Transpersonal Psychology 8(1): 45-64.

Tart CT (1989). States of consciousness. Psychological Processes, El Cerrito, CA.

Tart CT (1992). Transpersonal psychologies, American Psychological Association, Washington DC.

Taylor, E. I. (1992). Transpersonal psychology: Its several virtues. *The Humanistic Psychologist, 20*, 285-300.

Tornell H. (2001), the quantum man.

Thorndike, L. (1958). A history of magic and experimental science (Vol. 3). Columbia University Press

Turnbull, G. K., & Ritvo, P. G. (1992). Anal sphincter biofeedback relaxation treatment for women with intractable constipation symptoms. Diseases of the colon & rectum, 35(6), 530-536.

Turow, G. (2005). Auditory Driving as a Ritual Technology—A Review and Analysis. Unpublished Religious Studies Honors Thesis, Stanford University, Stanford, CA.

Ullman, M. (2003). Dream telepathy: experimental and clinical findings. In Psychoanalysis and the paranormal: darkness off lands (15-46 pp.). London karnac books.

Vallbona, C., & Richards, T. (1999). Evolution of magnetic therapy from alternative to traditional medicine. Physical medicine and rehabilitation clinics of North America, 10(3), 729-754.

Valverde, R. (2015) a. Neurotechnology as a Tool for Inducing and Measuring Altered States of Consciousness in Transpersonal Psychotherapy. NeuroQuantology, 13(4).: 502-517

Valverde R (2015) b. Channeling as an Altered State of Consciousness in Transpersonal Psychology Therapy. Journal of Consciousness Exploration & Research; 6(7): 405-416.

Valverde R (2011). Principles of Human Computer Interaction Design. Lambert Academic Publishing, Germany.

Vaitl, D., Birbaumer, N., Gruzelier, J., Jamieson, G. A., Kotchoubey, B., Kübler, A. & Weiss, T. (2005). Psychobiology of altered states of consciousness. Psychological bulletin, 131(1), 98.

Van Iersel, H., Bradley, K., Coon, D., Kendall, K., Stone, A., & Swaminathan, V. (2013). Nelson Psychology VCE Units 3 & 4.

Veitch, J. A., Gifford, R., & Hine, D. W. (1991). Demand characteristics and full spectrum lighting effects on performance and mood. Journal of Environmental Psychology, 11(1), 87-95.

Watkin, EI (1920), The philosophy of mysticism, London: G. Richards

Wallace B and Fisher LE (1991). Consciousness and behavior (3rd ed.). Allyn & Bacon, Boston.

Walter, W. G. (1964). Slow potential waves in the human brain associated with expectancy, attention and decision. European Archives of Psychiatry and Clinical Neuroscience, 206(3), 309-322.

Walter, V. J., & Walter, W. G. (1949). The central effects of rhythmic sensory stimulation. Electroencephalography and clinical neurophysiology, 1(1), 57-86.

Walsh R. N., & Vaughan, F. (Eds.). (1980). *Beyond ego: Transpersonal dimensions in psychology.* Los Angeles: Tarcher.

Watson, JB (1916). The place of the conditioned reflex in psychology. Psychological Review , 23 (2), 89.

Wiederhold, B. K., & Wiederhold, M. D. (2004). 14 The future of Cybertherapy: Improved options with advanced technologies.

Wilber K. Psychologia Perennis (1975): The Spectrum of Consciousness. Journal of Transpersonal Psychology 7(2):105-132.

Wilber, K. (1996). The Atman Project: A Transpersonal View of Human Development. Quest Books.

Williams, J., Ramaswamy, D., & Oulhaj, A. (2006). 10 Hz flicker improves recognition memory in older people. BMC neuroscience, 7(1), 21.

Williams, J. H. (2001). Frequency-specific effects of flicker on recognition memory. Neuroscience, 104(2), 283-286.

Wise, A. (1995). The High-Performance Mind: Mastering Brainwaves for Insight, Healing, and Creativity. GP Putnam's Sons, 200 Madison Avenue, New York, NY 10016.

Wickramasekera I, I. E. (1977). On attempts to modify hypnotic susceptibility: Some psychophysiological procedures and promising directions. Annals of the New York Academy of Sciences, 296, 143-153

Wright, A. J. (2011). Conducting psychological assessment: A guide for practitioners. Hoboken, NJ: John Wiley & Sons.

Wolberg, L. R. (1977). The technique of psychotherapy (Vol. 2). Grune & Stratton.

Wundt, W. (1980). Outlines of psychology (pp. 179-195). Springer US

Zusne, L. & Jones, W. H. (1982). Anomalistic psychology: A study of extraordinary phenomena of behavior and experience. Hillsdale, NJ: Erlbaum.

www.ingramcontent.com/pod-product-compliance
Lightning Source LLC
Chambersburg PA
CBHW022101170526
45157CB00004B/1425